Electric Choices

The **INDEPENDENT INSTITUTE**

THE INDEPENDENT INSTITUTE is a nonprofit, nonpartisan, scholarly research and educational organization that sponsors comprehensive studies of the political economy of critical social and economic issues.

    The politicization of decision making in society has too often confined public debate to the narrow reconsideration of existing policies. Given the prevailing influence of partisan interests, little social innovation has occurred. In order to understand both the nature of and possible solutions to major public issues, The Independent Institute's program adheres to the highest standards of independent inquiry and is pursued regardless of political or social biases and conventions. The resulting studies are widely distributed as books and other publications, and are publicly debated through numerous conference and media programs. Through this uncommon independence, depth, and clarity, The Independent Institute expands the frontiers of our knowledge, redefines the debate over public issues, and fosters new and effective directions for government reform.

THE INDEPENDENT INSTITUTE
100 Swan Way, Oakland, California 94621-1428, U.S.A.
Telephone: 510-632-1366 • Facsimile: 510-568-6040
E-mail: info@independent.org • Website: www.independent.org

Electric Choices

Deregulation and the Future of Electric Power

Edited by Andrew N. Kleit
Foreword by Pat Wood, III

Published in cooperation with The Independent Institute
Oakland, California
ROWMAN & LITTLEFIELD PUBLISHERS, INC.
Lanham • Boulder • New York • Toronto • Plymouth, UK

ROWMAN & LITTLEFIELD PUBLISHERS, INC.

Published in the United States of America
by Rowman & Littlefield Publishers, Inc.
A wholly owned subsidiary of The Rowman & Littlefield Publishing Group, Inc.
4501 Forbes Boulevard, Suite 200, Lanham, Maryland 20706
www.rowmanlittlefield.com

Estover Road
Plymouth PL67PY
United Kingdom

Copyright © 2007 by Rowman & Littlefield Publishers, Inc.

Published in cooperation with The Independent Institute

British Library Cataloguing in Publication Information Available

Library of Congress Cataloging-in-Publication Data

Electric choices : deregulation and the future of electric power / edited by Andrew N.
Kleit.
    p. cm.
  ISBN-13: 978-0-7425-4875-6 (cloth : alk. paper)
  ISBN-10: 0-7425-4875-9 (cloth : alk. paper)
  ISBN-13: 978-0-7425-4876-3 (pbk. : alk. paper)
  ISBN-10: 0-7425-4876-7 (pbk. : alk. paper)
  1. Electric utilities—Deregulation—United States. 2. Electric utilities—Deregulation—
Alberta. 3. Electric power—United States. 4. Electric power—Alberta. I. Kleit,
Andrew N.

  HD9685.U5E522 2006
  333.793'20973—dc22   2006012207

Printed in the United States of America

♾™ The paper used in this publication meets the minimum requirements of American
National Standard for Information Sciences—Permanence of Paper for Printed Library
Materials, ANSI/NISO Z39.48-1992.

Contents

Foreword

*Pat Wood, III, former Chairman of
the Federal Energy Regulatory Commission*

Americans know that market competition is good for consumers. It isn't just theory; it's our experience. The move to more market-based systems in many of our industries has been an important force in the United States' robust economic performance in the last two decades. This transition has been widespread, touching virtually every regulated industry: local and long-distance telecommunications, airlines, natural gas and oil transportation, trucking, rail roads, financial institutions, and more.

In my career as a utility regulator, I have witnessed how this wave of market restructuring has played out in the natural gas, trucking and telecommunications industries. And this economic transformation is finally coming to one of the last of the great twentieth century monopolies—the electric power industry.

For numerous reasons, many of which are explained in *Electric Choices*, the introduction of competitive market forces to the electric power industry has taken the longest. To be sure, there are physical and operational differences between the power industry and the others, but I have concluded that these differences affect *how* we should deregulate, not *whether* we should deregulate. What is most germane is that electricity is the most important commodity in our economy, underpinning virtually every aspect of our way of life. The gains from deregulation and restructuring are therefore high. But so are the potential risks. Restructuring the electric power industry therefore has drawn great scrutiny. This is as it should be.

As the son of a retail businessman, I tend to view most issues from a customer perspective. In this regard, my general view of restructuring the quarter-trillion dollar power industry has been driven by the answer to the question: are customers going to be better off over the long-term with the industry restructured along more market-based lines? Both despite and because of my fourteen years in utility regulation, the answer remains: Yes. More so now than ever.

The California energy crisis of 2000; the collapse of Enron; the significant run-up in natural gas, oil, and coal prices; the Northeastern North American Blackout of 2003—these events could have pulled me in the direction of saying, "No, electricity restructuring is just not worth it." But these significant events *reinforced* the need to finish the job and adopt a more transparent market model.

During this past decade as a regulator, I have learned some key lessons in how to disentangle government from an unhealthy involvement in a key economic sector. Here are a few.

GET THE SEQUENCE RIGHT

A certain former Governor of Texas once told me, "competition first, then deregulation." In my earliest days, the "small government" side of me would have replied, Deregulate, period! It didn't take me many months of experience to conclude that the Governor was right.

Technology and entrepreneurial creativity have transformed the fundamental economics of many industries. Industries like electricity generation, which in the early twentieth century were dominated by economies of scope and scale and thus experienced problems of natural monopoly, are today open to competition. Technological and entrepreneurial change prepared the way for removal of legal barriers to entry and eventual deregulation. But competitors don't arrive on the scene immediately after deregulation; workable competition develops slowly. In unraveling a century of regulation, it is important to establish a market first to provide the foundation for a successful transition. So called "deregulation" schemes that repackage regulated monopolies as market monopolies don't offer much to consumers.

At the same time it's important that regulators not become captured by today's dominant producers who fear creative destruction. As David Dismukes notes in chapter 9 the process of technological and entrepreneurial creativity is continuing in the electric industry—one day we may all own efficient electric generators (perhaps even in our garages!). Super-small generators would forever remove many of the problems of transmission that continue to bedevil deregulation. We need to make sure that preventing market abuse today does not preclude new markets tomorrow.

UNDERSTAND WHERE COMPETITION MAKES SENSE

The introduction of competition into network industries (energy and telecommunications) has been more complex and has taken a different course in these two backbone industries. We do not yet know how to create real competition in the transmission portion of the energy sector (but see William Hogan in chapter 7 for some insights). Deregulation is still worthwhile, however, because in energy, the network (pipelines, power lines and other components of the distribution infrastructure) accounts for no more than one-fourth of the end-user's energy bill. The rest of the bill reflects the costs of the competitive commodity that moves over the network. So, it is worth the hassle of "competition first, then deregulation" to get the competitive benefits on three-fourths of the customer's bill.

FIGURE OUT WHAT IS NECESSARY AND DO IT WELL

Because energy networks are the enablers of competition, state and federal regulators of the energy industry have focused their efforts on opening up access to the grid. FERC Order Nos. 436, 500, and 636 and similar efforts at the state local distribution company level have opened up the gas network, and FERC Order Nos. 888, 2000, 2003, and 2006, together with similar state efforts on the distribution systems have removed barriers to entry on the power grid.

Opening up the delivery systems to competing users was the critical step. Utilizing traditional antitrust tenets that prohibit anticompetitive practices on essential facilities, regulators have enabled competition in the provision of natural gas and power. In most cases, reducing or even eliminating the ability of incumbents to exert vertical market power has led to a natural restructuring of the energy industry along functional lines. And this natural course of events has led to an efficient and rational investment in infrastructure that supports market development and serves customers well.

As a regulator, it has been tempting to move from the necessary to the desirable and impose more specificity on the restructuring effort. But if history has shown anything, it has shown that the future is going to differ greatly from what anyone plans. Exercising restraint and focusing on only the necessary tasks is therefore essential.

KEEP AN EYE ON THE BALL—AND THE BALLGAME

It is crucial to stay focused on the long-term goal of restructuring: customer benefits. These industrial transformations take time. Government administrations and regulators come and go, but it is important that the direction, once adopted, be assured. There have been times when I haven't agreed with previous regulatory decisions, but for the sake of continuity and a consistent direction, I took them as given and built upon them. In industries where many billions of dollars of investments are made every year, even a mediocre decision that is upheld is preferable to an excellent decision that breaks faith with what came before.

As a separate aspect of keeping an eye on what matters, vigilant oversight of nascent markets is necessary. Watchful monitoring of anticompetitive practices, market failures, and market power is required to ensure that the hard-fought customer gains are converted into tangible value. Protecting more vulnerable customers from abusive practices or from the extreme price volatility of commodity markets is also important, although such protection must be done in ways that don't undermine the clarity of market signals. Getting this balance right has been a challenge.

LET FREEDOM RING

With competition come the traditional first-tier benefits: better pricing than under regulation and more responsive (and tailored) service for customers. But even more important than these first-tier benefits are the benefits that we cannot predict. Deregulation unleashes technological innovation. We have seen this in the natural gas production, power generation, and telecommunications arenas in the last quarter century. In chapter 3 Lynne Kiesling discusses how we are beginning to see innovation on the customer side in the electricity industry as the monopoly over the meter is eliminated. We are even seeing spillover effects with advanced technology in the energy delivery businesses even as these remain regulated. Another positive effect of competitive markets is the resulting emissions cleanup of the power plant fleet, most notably in my home state of Texas.

Customer choice—good ol' "Made in the USA" freedom—is driving businesses to innovate and excel in a way no government regulator ever could. In that light, I hope you will relish the following lessons about the past and future of the electricity industry from some of the brightest people I have gotten to know over the past decade. The superb and very timely analysis and recommendations in *Electric Choices* will help us to finish the revolution. Let's win this one for the customer.

1

Introduction

Andrew N. Kleit

Since the late 1970s, the advance of the economic idea of free markets has dramatically altered the regulatory landscape in the United States. Airlines, railroads, trucking, and other areas were deregulated, with impressive economic results.

With success in so many other areas, it is not surprising that the liberalization movement reached the electricity industry. The drive for restructuring electricity markets was pushed by several factors. Intellectually, perhaps the most important event was the 1978 passage of the Public Utilities and Regulatory Policy Act (PURPA), which allowed independent generators to sell their electricity to utilities at rates often determined by avoided costs (in essence, marginal costs). PURPA allowed small electricity generators to sprout up around the country. In doing so, it demolished the intellectual idea that electricity generation was a natural monopoly that required regulated prices and entry.

In the purely political area, probably the largest impetus for restructuring was the massive cost overruns on nuclear power plant construction. In the 1960s through the 1980s a large number of nuclear power plants were built around the country. These plants were built at tremendous cost overruns. When the bills came due, in the tens and perhaps hundreds of billions of dollars, consumers started to complain, and politicians started to notice.

Today we refer to those cost overruns as the "stranded cost" problem, that is, utility costs that cannot be recouped in the marketplace. Why did the stranded cost problem occur? Who was at fault? Who should pay for it? In retrospect, it appears that the stranded cost problem was the fault of many. Electric utilities, ignoring the law of demand, assumed that the demand for their product would continue to rise even as its price increased. The federal

1

government passed laws precluding the development of new fossil fuel plants, and then tightened regulations on nuclear power plants. Regulatory commissions across the country approved as "prudent," that is, worthy of being repaid by ratepayers, expensive construction that should have been disallowed. In retrospect, the tremendous waste of money seems almost inevitable.

Why did this waste occur? The answer is simple. Everyone thought he was playing with someone else's money, creating the results to be expected from such poor incentives. It seemed clear that if, in the end, the nuclear power plants did not work, the regulators would not pay the bill; the federal government would not pay the bill; and most of all, the electric utilities and their shareholders would not pay the bill. In the end, the consumer would pay the bill. Indeed, given a fixed rate of return, the higher the cost overruns, the higher the profits.

Restructuring is one way to make sure that the stranded cost debacle does not happen again. In a restructured electricity sector, with wholesale power sales deregulated, investments in power plants are just as risky as investments in any other nonregulated sector of the economy. If we have another round of overly expensive power plants with limited demand for their products, it will be investors who pay. So the consumer, at least in restructured states, is now protected from a repeat of the stranded cost problem.

While restructuring may well solve any future stranded cost problem, it is not a panacea. The prospect of restructuring leaves us with many problems, and many opportunities. The opportunities in large part lie in marketing electricity to consumers. Historically, consumers have been offered a fixed rate for power, no matter what time of day they used such power. Since the cost of power fluctuates wildly during the day, the chance to offer differential pricing presents the opportunity for large efficiency gains. Other gains are available in product differentiation in electricity, as well as structural reform in such areas as transmission and system operation.

The momentum behind avoiding more stranded costs, opening up new retail opportunities, and a general wave toward deregulation led to restructuring passing in Pennsylvania, California, Texas, and a variety of other, mostly northeastern, states. At the same time, resistance to restructuring was strong in many areas, especially in the South.

The electricity debacle in California in 2000–2001 caused that tide to come to a crashing halt. A combination of poor institutional design, adverse economic factors, and political actors taking predictable courses caused the electricity market to collapse in that state. The events in California have, in large part, taken the steam out of the deregulatory movement, at least in electricity.

That leaves the United States with an electricity market that, at the retail level, is half deregulated and half regulated. States such as Pennsylvania,

New York, and Texas have engaged in electricity restructuring. Meanwhile, other states, such as Florida and Minnesota, appear to have no desire to move in this direction. On a national scale, it is not obvious whether restructuring will emerge as the dominant form in electricity regulation.

Restructuring electricity is, in a word, difficult. It is difficult for several reasons. As Hunt (2002, 29–33) points out, electricity is not storable, it follows the path of least resistance across the transmission grid, the transmission of power is subject to a complex series of far-reaching interactions, and electricity travels at the speed of light. All this implies the need for short-term (instantaneous) coordination in electricity markets, requiring some regulatory structure to remain in place in even the most liberated market regime.

Thus, it is clear that there must be some type of system coordinator to balance supply and demand in a control region. It is also clear that, at least for the time being, a single distributor of electricity is required, at least for most customers. What is not clear, as chapters in this volume make clear, is whether coordination and distribution should be done publicly or privately and if privately by profits or nonprofits.

Each chapter in this volume deals with an important issue in restructured electricity markets. Chapter 2, by Considine and Kleit, reviews restructuring efforts in California and Pennsylvania. They also estimate the composition of factors contributing to higher electricity prices in 2000–2001 in California after restructuring. Considine and Kleit find that even if California had not deregulated its electricity industry, it would have faced higher electricity prices for several reasons, including a shortage of generation capacity and bottlenecks in producing and delivering additional natural gas supplies for power generators.

By comparing the two regulatory regimes, Considine and Kleit come up with several recommendations for restructured electricity markets. First, restructuring should not place any limits on trading in wholesale markets. There is no reason that consumers in electricity markets should be denied access to the market for risk, which exists in many other areas.

Second, Considine and Kleit also find substantial drawbacks to the use of price caps in both wholesale and retail markets. Retail price caps send the wrong signal to consumers. Electricity became scarce in California in the period 2000–2001. Consumers in California, however, had no incentive to reduce their consumption of electricity, as their prices were fixed. Fixed consumer prices served to both bankrupt utilities in the state and to place tremendous strain on the electrical grid.

While there was previously a debate over whether stranded cost recovery was appropriate, that debate is now over. An integral part of any restructuring plan is the recovery of utilities' stranded costs. Unfortunately, the method chosen for doing so in California and Pennsylvania was inappropriate. Considine and Kleit discuss whether a stranded cost "tax" might be attached to every kilowatt-hour of power sold in relevant areas. Consumers would then be free to contract for their power supply.

Lynne Kiesling examines the retail side of the picture in chapter 3. She contends that the "one size fits all" of regulated and fixed rates is becoming increasingly obsolete because of technological, institutional, regulatory, and cultural changes that recognize the diversity of products that the electricity industry can profitably sell to consumers. Without retail pricing reform, electricity restructuring will fail to deliver efficiency and value to consumers. Retail reform is thus necessary to maximize the value of other market reforms and is in and of itself a valuable step in producing an efficient electricity market.

Stephen Rassenti, Vernon Smith, and Bart Jones review the use of experimental economics to inform the debate on electricity market regulation in chapter 4, reprinted from the 2002 *Cato Journal*. The chapter reviews experimental studies on electricity restructuring from the mid-1980s. These studies demonstrated in practical terms how electricity markets could be divided into transmission, system operation, and marketing in a practical sense, laying the groundwork for practical restructuring in the 1990s.

In particular, experimental economics shows that the use of demand-side bidding for a relatively small fraction of the market, much like what Kiesling discusses in chapter 3, can serve to reduce price spikes and the exercise of market power. This implies that large gains are available from making demand prices available to consumers. In addition, demand-side bidding would ease the need (or the perceived need) for price caps in many regions. Similar experimental results also show how competitive transmission markets can be constructed.

Chapter 5, by Terry Daniel, Joseph Doucet, and André Plourde, examines an otherwise neglected part of the restructuring experience. The province of Alberta has successfully restructured its electricity sector. The Alberta market has most of the essential elements, and problems, of electrical systems elsewhere, but it is modest in size and somewhat isolated from many complicating factors that tend to confound analysis of larger more interconnected systems. As such it is an ideal test case for policy analysis. In addition, there are several aspects of the implementation process that have been chosen for Alberta that are unique, interesting, and worth considering.

In particular, unlike systems in the United States with stranded costs, the average cost of power supply in Alberta was actually less than the marginal cost of supply. Thus, rather than stranded costs, restructuring in Alberta threatened to transfer to the utilities "stranded benefits." This issue was dealt with in two successive steps in Alberta, namely the implementation of "legislated hedges" in the wholesale market and later, the auction of the energy produced from regulated generation capacity.

To deal with issues of market power, a controversial and unique element of the Alberta restructuring plan obliged the owners of the regulated generation units to sell the ownership rights to the *energy* production from the remaining (regulated) life of these plants at a one-time auction. The purchasers of these rights, possibly but not necessarily new players in the market, would then be responsible for bidding the energy into the Alberta Power Pool on a daily basis. In this sense, this one-time auction, with strict limits on purchases, would reduce market concentration of the energy output of regulated units. The hope was that a sizable collection of new participants would be attracted into the supply-side market, creating a new level of competition leading to lower electricity prices. The second goal of the sale of regulated energy was to capture the "stranded benefits" associated with the low-cost embedded generation. Daniel, Doucet, and Plourde evaluate the success of this regulatory innovation and subsequent market developments.

Chapter 6, by Craig Pirrong, asks: How far can decentralization go in power markets? He analyzes the costs and benefits of decentralization in power markets using the tools of transaction costs and property rights economics. The basic conclusion of his analysis is that complete decentralization may be infeasible in power markets, owing to the physical characteristics of power transmission systems. In particular, reliable operation of a power system may be a "public good" that creates well-known collective-action problems that the creation of a central coordinating agent can mitigate. Moreover, the time scale of power system operations can create acute holdup problems in decentralized contracting setups.

Transactions cost and property rights considerations also imply that the form of organization of the coordinating agent and the legal environment in which it operates have important efficiency implications. Under the circumstances present in the electricity industry, "low-powered" incentive regimes (such as profit regulation of transmission companies or the formation of nonprofit coordinating agents in which there is an attenuated relation between agents' compensation and their performance) are likely to be efficient.

Moreover, the governance structure of the coordinator matters. In a disintegrated power market, a nonprofit organization controlled by stakeholders who consume, produce, or market power (such as an independent system operator) is plausibly a form of organization that economizes on transactions cost. Financial exchanges have in fact adopted the nonprofit cooperative-stakeholder-controlled organizational form and thus may serve as a role model for power markets.

Another difficult issue for electricity restructuring is transmission. The common wisdom on transmission is that because it exhibits economies of scale, it is not subject to restructuring. William Hogan, the leading writer in this area, takes issue in chapter 7 with the common consensus. In his reworked 1999 study included here, he discusses how transmission markets can be at least partially restructured.

The core of his idea works something like this: Assume a transmission line running from points A to B. At point A the price of electricity is 4 cents per kilowatt-hour. At point B, the price is 5.5 cents per kilowatt-hour. Across this transmission line run 10,000 kWh of power. Given this situation, the transmission congestion "rent" is equal to the price different between A and 5, or 1.5 cents per kWh. The owners of the transmission line therefore receive the total congestion rent across their line, 1.5 cents per kWh times 10,000 kWh, or $150. Given this framework of property rights, Hogan goes on to show how transmission markets can move toward a competitive market. In particular, Hogan discusses how such a framework can induce efficient entry in the transmission sector.

To enhance entry into transmission markets, Hogan suggests that owners of new transmission be able to restrict access across their lines in order to capture monopoly profits. While this proposal is consistent with standard antitrust law, which allows a firm to restrict its own capacity, it is in conflict with Federal Energy Regulatory Commission (FERC) policy, which calls for open and complete access to such facilities.

In chapter 8, Timothy Brennan addresses an important issue: Can market power be measured in electricity? Any successful restructuring plan will seek to limit the available opportunities for firms to reduce output and therefore raise prices to consumers. When it comes to the empirical assessment of market power, however, Brennan asserts that the approach taken in most of the analyses of market power in electricity rests on a flawed application of a standard measure of market power—the Lerner index, also known as the price-cost margin. The fundamental rationale for using price-cost margins is essentially that in a competitive market, price-taking firms will supply output up to the point where the marginal cost of production just equals the market price. Therefore, a substantial difference between price and marginal cost indicates that firms are not taking price as given.

Brennan argues that the flaw in many electricity market studies is not that the price-cost margin is theoretically inappropriate, but is in the manner in which it has been implemented. In these studies, the proxy for "marginal cost" used to estimate price-cost margins is typically the average variable or operating cost of the "last" generator that would be dispatched to meet energy demand.

Erroneous use of price-cost margins to measure market power is not merely a matter of academic interest. Such measures could lead, and perhaps already have led, regulators to prevent sales of electricity above the highest average variable cost of the generators used to provide electricity. Keeping at least some firms from earning revenues in excess of their average variable costs will encourage present suppliers to leave the market and discourage new firms, particularly those needed to provide power on-peak, from entering. Without such entry, competitive electricity markets are likely to fail

The potential to open the retail market brings new opportunities in price stability, supply reliability, and capital investment, issues that may well be best addressed by the interaction of supply side and demand side, between the wholesale and retail levels, not the myopic supply approach of traditional regulation. To allow for static and dynamic efficiency in electricity markets, Kiesling contends that we need demand-response expression in markets, not simply isolated demand response, curtailment programs, and experiments, as have been offered by some utilities in the recent past.

The standard model of electricity competition is a relatively large generation plant, sending electricity through a natural monopoly distribution system to customers. What if, instead, a substantial number of customers could produce their own power? David Dismukes explores this possibility in chapter 9, outlining the prospects for distributed energy resources (DER).

DER refers to generation, storage, and demand-side management (DSM) devices and technologies that are connected to the electric grid at the distribution level (i.e., below the bulk power transmission system). In many instances, DER can provide the electricity consumer with greater reliability, higher power quality, and more flexible choices.

Today's computer server-farms, and a large portion of high-tech manufacturing, require much greater reliability and quality than what has traditionally been delivered. In addition to power quality benefits, many DER technologies can be more environmentally friendly and less expensive than the provision of typical utility service. Widespread use of DER technologies could also mitigate congestion in transmission lines, help to control price fluctuations, and provide greater stability to the electricity grid. Distributed energy resources, however, threaten traditional utilities and will not be widely adopted unless regulatory changes are made so that DER can fit into the existing paradigm.

One of the most important advances in economics in recent years has been the use of experimental tools to test economic theories. The importance of this area was acknowledged in Vernon Smith's receipt of the Nobel Prize for Economics in 2002. Experimental economics allows for the testing of a variety of economic hypotheses in a laboratory setting, saving economics from much sheer speculation.

In the book's final chapter, Lynne Kiesling and Michael Giberson review the U.S.-Canada Power System Outage Task Force on the northeast American blackout of August 14, 2003. Electric power blackouts of any magnitude bring the system's reliability into question. The sheer size of the northeast American blackout of August 2003, affecting more than fifty million customers, has prompted a lengthy and ongoing examination of the transmission grid. Understanding the causes of the blackout was an important first step to setting a policy course for a reliable twenty-first-century grid. Unfortunately, the blackout report suggests policies based almost strictly on an engineering and regulatory point of view. The primary ultimate impact of the forty-six recommendations would be the expansion of regulatory oversight over supply-side reliability decision making. Such policies would not enhance system resiliency, flexibility, and adaptability. To achieve those objectives, a more distributed and decentralized approach is needed.

Kiesling and Giberson suggest a more economic approach would be better advised. They suggest a more resilient, flexible, dynamic transmission system, a transmission grid that better adapts to the demands that are placed upon it and to unknown and changing conditions. In particular, they assert that the most important changes to make in the industry are really just a continuation of industry restructuring. They focus on the possibility of commercializing reliability by reforming the reliability rules to properly line up incentives and information flows. As they point out, reliability is valuable to consumers. What has been lacking is a way for consumers to express that value, and for suppliers to be paid appropriately for providing it.

Properly restructuring electricity markets is a difficult and challenging problem. But it is not going away. There seems to be no doubt that the generation and marketing of electric power are not a natural monopoly. Thus, regulators need to decide how much they are going to eliminate restrictions on these markets. In addition, there are further opportunities for market advances in the distribution, transmission, and system operation levels of the electricity industry. There is no apparent reason to retain the industry monopoly structures in these areas.

The virtue of market forces can play a large role in bringing about a more efficient electricity sector. But electricity restructuring requires careful policy choices. The chapters in this volume lay out various means for reaching more competitive and better-operating electricity markets.

REFERENCES

Hunt, Sally. 2002. *Making Competition Work in Electricity*. New York: Wiley.

2

Can Electricity Restructuring Survive?
Lessons from California and Pennsylvania

Timothy J. Considine and Andrew N. Kleit[1]

For almost a century, electricity was a tightly regulated industry, from generation to transmission to distribution. Starting in the mid-1990s, however, reform efforts began in several states to open up electricity markets to the forces of competition. The reformers brought with them promises of lower prices and more efficient operation. Today, however, there are many questions about efforts to restructure the electricity industry.

Nowhere have concerns been more pressing than in California. With soaring electricity prices, blackouts, and bankrupt utility companies, the economic and political fabric supporting regulatory reform of the electricity industry in California has been severely frayed. Based upon this experience, many states are hesitating or delaying plans to deregulate their electricity industries. Pennsylvania, however, has progressed down the path of reform with no apparent deleterious effects. This chapter attempts to understand why there is such a sharp dichotomy in these experiences, and what can be learned from them.

The seeds of deregulation were sown in 1978 when Congress passed the Public Utilities and Regulatory Policy Act (PURPA), which allowed independent generators to sell their electricity to utilities at rates often determined by avoided costs (in essence, marginal costs). Also during this time, new electric power generation technology using jet engines recycling exhaust gases became commercially viable, operating efficiently at relatively small scale with clean and relatively inexpensive natural gas. Indeed, most new generation capacity in the United States since 1985 has come from independent power producers using combined-cycle natural gas turbine technology.

During the early 1990s it became clear that fundamental change was occurring in the generation of electric power and that regulatory policy needed to respond. Congress acted in 1992 by passing the Energy Policy Act. This induced the Federal Energy Regulatory Commission (FERC) to issue Order 888 in 1996, forcing utilities with transmission networks to deliver power to third parties at nondiscriminatory cost-based rates.

With this door open, several state public utility commissions adopted customer choice programs and other policies to disaggregate retail prices into generation, transmission, distribution, and transition charges (Joskow, 1996). Unlike the deregulation of many other industries in which government involvement was phased out completely, however, deregulation of electricity is more complicated. In many ways, electricity markets are not really deregulated but restructured.

Our analysis first compares restructuring plans in California and Pennsylvania. We also examine the performance of electricity markets in California and Pennsylvania after restructuring. In particular, we estimate the composition of factors contributing to higher electricity prices in 2000–2001 in California after restructuring. We find that even if California had not deregulated its electricity industry, it would have faced higher electricity prices for several reasons, including a shortage of generation capacity and bottlenecks in producing and delivering additional natural gas supplies for power generators.

Our comparative analysis of restructuring efforts in California and Pennsylvania raises a number of broader regulatory policy issues that are addressed below. These issues include monitoring markets for excessive markups over marginal cost, marketing, and some rather thorny issues involving market coordination between generators and transmitters of electric power.

WHAT IS ELECTRICITY RESTRUCTURING?

Deregulation generally entails the removal of some form of overt government controls on decisions by firms. Until recently, nearly all investor-owned electric utilities in the United States were subject to rate-of-return regulation in which companies petition state public utility commissions for permission to charge customers rates that would cover expenses and a rate of return to stock- and bondholders. Most of these companies were vertically integrated, owning generating plants, intercity transmission lines, and local distribution networks. Eliminating rate-of-return regulation generates a number of potential benefits but also creates the need for a market to determine electricity rates. *Restructuring* essentially involves the creation of these markets and the dissolution of the vertically integrated structure of the industry.

Our discussion begins with a review of the potential gains from restructuring. It then presents an overview of the basic economics of electric power generation, transmission, and distribution. Given these basic economic features, we then develop the rationale for regulation and restructuring in the industry. From this context, we then describe the general forms of deregulation.

THE POTENTIAL GAINS FROM RESTRUCTURING

There are two particular areas in electricity markets where it is hoped restructuring will bring about significant economic gains. First, restructuring frees electricity generators from rate-of-return regulation and fuel adjustment clauses. Generators thus have important incentives to cut costs, which will result in lower prices for consumers in the long run. No longer able to pass fuel costs on to customers via fuel adjustment clauses, electric generating companies would take a much tougher stance in negotiating contracts for coal, natural gas, and other fuel sources. In addition, power generation firms would be forced to take a similar approach in their negotiations with hourly and salaried employees. Considine (2000) estimates that restructuring the electric power generated by fossil-fuel-fired steam alone would lower annual long-run costs by $7 billion due to increased efficiency of operations, achieved by a more efficient use of capital, fuel, labor, and maintenance. In areas with excess capacity, these lower costs in a competitive market will naturally decrease the price of power.

Second, restructuring eliminates the monopoly on retailing held by local distribution companies. In a properly restructured market, any number of providers can compete on both price and quality of service when offering retail electricity to consumers. Consumers could choose their electricity producer just as they select their long-distance telephone provider. This competitive environment could lower rates and improve the quality of service. Moreover, competition would improve the efficiency of allocating electricity because a competitive structure matches consumer preferences with production realities.

BASIC PHYSICAL FEATURES OF ELECTRICITY SUPPLY

The generation and provision of electricity involves a network of generators and high-voltage transmission lines connecting these generators with transformers that reduce the voltage of the electric power for distribution to final consumers. Once the power has been generated, it is sent over transmission lines to the relevant market. Transmission lines are often expensive and must travel through a variety of areas. One difficulty with building new transmission lines is that many communities do not wish to have them in their area. This makes obtaining regulatory approval for these lines difficult.

Power comes through transmission lines into a particular "control area." Electricity cannot be stored (at least inexpensively), and thus there is a requirement that the demand for power in a control area be set equal to the supply. In addition, power flows must be controlled so that a variety of system requirements are met. Historically, each relatively large utility controlled its own "system operations." In recent years, however, several independent system operators have arisen to control systems spanning across several utilities.

Once power comes into a particular control area it is dispatched through lower-density distribution lines to consumers' homes and businesses. Generally speaking, production costs appear to be minimized if only one distribution line exists to particular customers.

RATIONALE FOR REGULATION AND RESTRUCTURING

Economic theory teaches that if the right conditions are in place, unrestricted entry and price competition will create the outcome that creates the most economic wealth for the relevant society. One important condition for a viable free market is a sufficient number of viable independent competitors. This, in turn, implies that the total cost of supplying customers is not minimized by having only one firm in the market, that is, that the market is not a "natural monopoly." Indeed, the apparent rationale for the regulation of electricity, beginning in the early years of the twentieth century, implied that competition was not viable in this industry. Today, the distribution of power appears to fit this natural monopoly criteria, as well as system operation. Transmission lines, at least in some areas, may also fit these criteria.

While it may be the case that distribution, system operation, and perhaps transmission are in some sense "natural monopolies," the same cannot be said about the generation of electricity. In most significant economic regions, electricity is generated at a large number of locations. In addition, there is no obvious reason why the retailing of electricity constitutes a natural monopoly.

The potential competitive nature of the generation of electricity was revealed by the implementation of the Public Utilities and Regulatory Policy Act of 1978 (PURPA). PURPA mandated that independent producers of electricity have access to the regulated utility grid. Dismukes and Kleit (1999) find that PURPA increased the number of electricity producers, in particular, producers not affiliated with local regulated electricity providers.

GENERAL FORMS OF DEREGULATION

Restructuring calls for customers to obtain access to competitively generated electricity through monopoly distribution and transmission lines. There are two basic methods that have been used to grant such access. In the first arrangement, the price of power is determined on an exchange, and that price is passed along to consumers. This was the approach originally used in California. The second is "direct access," in which consumers can either buy power on an exchange or with a contract. This is the approach used in the Pennsylvania-New Jersey-Maryland (PJM) market.

Even with restructuring, electricity supply must equal demand and the system must be reliable. In several markets, an Independent System Operator (ISO) or similar Regional Transmission Organization (RTO) performs this task. Questions remain, however, about the appropriate scope of an ISO's powers, and the appropriate corporate governance structure for ISOs. We discuss some of these issues below. Transmission pricing is another source of debate. Competition in the transmission of electricity is difficult for two reasons. First, economies of scale in transmission may preclude effective competition in that area. Second, as Chao and Peck (1996) among others have illustrated, electrical transmission is subject to "loop flow," where one party's transmission decisions can affect another party's transmission capacity.

CREATING NEW MARKETS

Restructuring has created new profit opportunities for the trading of electricity rights. In particular, trading of electricity futures has also increased. In an electricity futures market, a buyer of power pays a certain amount of money for the delivery of a specified amount of power at a specified date in the future. In this manner, a buyer can insure itself against fluctuations in the price of power. Future markets in a variety of commodities are used as a method of reducing risk, and therefore increasing consumer satisfaction. Unfortunately, as we will discuss below, the original California restructuring plan prohibited the exchange of futures contracts.

In regulated states, consumers generally pay a constant price for power, no matter what the wholesale price of power is. This can generate serious economic inefficiencies. While the retail price of power paid by a consumer may be constant across a particular day, the marginal cost of producing that power can vary widely, as demand fluctuates. The result is that too much power is consumed during high-demand periods, and too little power is consumed during low-demand periods.

Restructuring may allow consumers to pay a price for power that is a function of its contemporaneous wholesale price. If these consumers could shift their electrical consumption to low-demand periods, they could gain from such "time-of-day" pricing. Time-of-day pricing may be especially advantageous to firms that can shift their production activities to the nighttime or to weekends.

APPROACHES TO RESTRUCTURING

There is no one method to undergo electricity restructuring. In this section, we examine the restructuring decisions made in California and Pennsylvania.

California

In March 1998 the state of California began its experience with restructuring, and wholesale electricity prices were no longer regulated. In order to alleviate concerns about the exercise of market power, the privately owned utilities were required to sell 50 percent of their generating capacity. These utilities eventually decided to sell 100 percent of their power, though much of it was sold to their own unregulated subsidiaries. This left the former major utilities in California, at least their regulated elements, as simply regulated distribution companies, subject to traditional rate-of-return regulation.

Utilities in California had made significant investments in nuclear power, investments that would be nonremunerative in a restructured world, or producing "stranded costs." Stranded costs are investment made by private utilities that would be nonremunerative in a competitive market. The state restructuring plan allowed for these stranded costs to be paid by a state-mandated fund. The fund gained its monies by the difference between the allowed retail price, and the appropriate expenses, such as wholesale power acquisition, transmission and distribution fees, and other charges incurred by the relevant distribution company. The amount of stranded cost recovery allowed each utility was determined administratively by the state Public Utility Commission (PUC).

All power that was to be sold to final consumers in the state was required to be sold into a power exchange operated by the state. No long-term contracts were allowed. Thus, producers of power were forced to sell their products in the hourly spot market, and consumers were forced to buy at this price.

This requirement was eliminated in August 2000, as the wholesale price of power in California surged. The power exchange was disbanded in January 2001. In addition, the state of California spent more than $1 billion to create its own Independent System Operator. This ISO served to combine the control areas of the three major privately owned utilities in California.

Individual retail customers were given the option of finding their own sources of electricity. However, the system was designed so that the discount any consumers received from their distribution company for choosing their power supplier was set equal to the power pool price. This implied that customers could not receive lower prices by choosing their own suppliers. Once this plan became known, most retail power suppliers left the market.

With the onset of restructuring, retail prices in California were reduced 10 percent. Retail prices were fixed at this level until, at least according to the original plan, stranded costs were paid off. Once stranded costs were paid off, the marketplace would have determined the price of power. Presumably, at this point, other retail suppliers would have found it economical to enter the market. Note that this system implied that prior to the full payment of stranded costs, retail prices would not be a function of wholesale prices. In the period 2000–2001, this would prove to be a serious mistake.

Independent power producers are allowed to create their own electricity generating plants in the state of California. Such plants, however, must meet local zoning and environmental regulations. As prices increased in the year 2000, many observers commented that such regulations in California were far too strict, contributing to the electricity shortfall in the state.

Pennsylvania

Electricity restructuring in Pennsylvania was phased in during the period from July 1998 to January 2000. Several small cooperative utilities in Pennsylvania were unaffected by this change. Electricity generators were set free to charge whatever price the market would bear for their product. No divestitures were required, as the state Public Utility Commission determined that there would be no significant market power problems upon restructuring. Several utilities independently chose to sell many or all of their generation facilities.

As in California, arrangements were made to pay off electric utilities' stranded costs. The value of these stranded costs was determined in administrative proceedings. Funds equal to the difference between the retail price of power, and the costs of transmission, distribution, and the price of wholesale power were used to account for stranded costs. As in California, the retail price of power to consumers was fixed (for those consumers who did not choose an alternative supplier) and, therefore, independent of fluctuations in the wholesale price of power.

Upon passage of the restructuring bill in 1996, electricity rates were frozen across the state. The state PUC then entered into restructuring agreements with several local utilities that reduced power rates. In addition, utilities no longer had the right to pass along their increases in fuel costs to customers. As was the case in California, these rate freezes are planned to continue until stranded costs are paid off.

Consumers who chose to pick an independent supplier were given a "shopping credit." These credits reduce consumer charges from their local distribution company. The value of the shopping credit is designed to proxy the value of power that these consumers are no longer buying from their distribution company. In perhaps simpler terms, consumers benefit if they can buy power from independent suppliers at a lower rate than the shopping credit.

The shopping credit is set administratively by the state PUC for each local distribution company, and was initially designed to be slightly above the wholesale price of electricity. This was meant to give an inducement to retail suppliers to enter the market. Unfortunately, the shopping credit does not fluctuate with market prices. During the 2000–2001 period, with higher natural gas prices, the wholesale price of power rose above the relevant shopping credit level, and most retail suppliers withdrew from the Pennsylvania market. By early 2005, they had not returned.

The state of Pennsylvania did not set up its own system operator to merge the system operations of utilities. In the eastern part of the state, the utilities had already pooled their system resources into the PJM system operator. In the western part of the state, utilities continued coordinating their own systems. In 2003, however, these utilities entered into an extension of PJM, often referred to as "PJM-West."

Independent power producers were given the right to access the power grid with their own production, and many have done so. The chief impediment here does not appear to be local zoning regulations, though independent producers must address these issues. Rather, it appears that access to the system operator can be difficult to obtain.

MARKET PERFORMANCE UNDER RESTRUCTURING

The problems in California under electricity restructuring are well known; bankrupt utility companies, blackouts, and sharply higher prices. In contrast, the restructured electricity market in Pennsylvania has enjoyed reliable supply at relatively low and stable prices. The objective of this section is to understand the reasons for such radically different outcomes under restructuring. Our principal finding is that the higher prices in California were inevitable due to a shortage of electricity production capability and a heavy reliance on power generation fired by natural gas. Blackouts and many of the financial difficulties resulting from the market meltdown in California, however, were due to fatal flaws in the restructuring provisions.

Another important finding is that Pennsylvania is much less dependent on natural gas than California. The experience of California suggests that price volatility may be one consequence of this reliance on natural gas. More generally, policy makers considering restructuring electricity markets, which we believe will generate long-term benefits to society, should consider the role of environmental regulations in achieving a diversified portfolio of electric power generators.

The discussion in this section begins with an overview of the performance of electricity markets under restructuring in California and Pennsylvania. A shortage of electricity capacity was a major factor behind the spike in electricity prices in California during late 2000 and into early 2001. Another major contributing factor was a dramatic increase in natural gas prices.

The California Experience

As the previous section explains, the California restructuring plan forced the three main electric utility companies in the state—Pacific Gas and Electric, Southern California Edison, and San Diego Gas and Electric—to sell half of their generation capacity. The restructuring plan required most power to be bought and sold in the California wholesale power exchange set up by the state. Also beginning January 1, 1998, residential customers of the investor-owned utilities received a 10 percent reduction in their monthly bills. Consumer rates include a distribution and transmission charge, a generation charge, other miscellaneous charges, and a competitive transition charge (CTC) that was used to pay off stranded costs. For example, a customer of Southern California Edison on average paid 12.7 cents per kilowatt-hour in 1999 (see table 2.1). More than 4.6 cents of that reflected a still-regulated transmission and distribution charge. The generation charge was approximately 3.2 cents. Other miscellaneous charges amounted to a shade more than 2.3 cents. The CTC picked up the remainder, 2.5 cents per kilowatt-hour. Under the plan, consumer rates were frozen until stranded costs were paid off. Note that the CTC charge is a residual equal to the fixed price to consumers minus transmission, distribution, and other charges, and minus the fluctuating generation price. As long as the generation price did not go "too high," this system was financially stable.

Table 2.1.   Average Electricity Rates for Southern California Edison Co., 1998–2001

Component	1998	1999	01/2000–04/2000	05/2000–02/2001
		Average Rate in Cents Per Kilowatt-hour		
Generation charge	3.34	3.22	3.78	17.36
Transmission & distribution	3.34	4.64	5.66	3.44
CTC	3.28	2.50	1.18	−10.11
Other charges	2.76	2.31	2.10	2.17
Amount paid per month	12.72	12.67	12.72	12.86

Under this scheme, generators have important incentives to cut costs, one of the two objectives of restructuring. In the California system, however, there was little room for retail competition. Any consumer who chose to purchase power from a retailer other than the local distribution company received a rebate equal to the wholesale price. This meant that a retailer could not show a profit unless it was able to purchase power below the wholesale price, a task which was close to impossible. For example, if the wholesale price was 3.5 cents per kilowatt-hour, 3.5 cents was the amount of the rebate. Since no one would sell to retailers at less than 3.5 cents when they could get this amount in the wholesale market, no electricity retailer could make money in California. Thus, the California system precluded retail competition until stranded costs were paid off.

Unfortunately, by the summer of 2000, the California system unraveled, as the average wholesale price of electricity rose more than tenfold. Figure 2.1 displays weekly average wholesale prices for electricity in the California Power

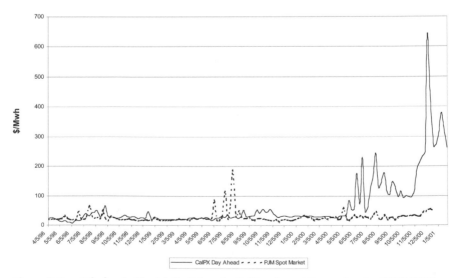

Figure 2.1.   Wholesale Electricity Prices in California and Pennsylvania, 1998–2001

Exchange (CalPX) and in the Pennsylvania-New Jersey-Maryland (PJM) market. Prices in California were relatively low and stable from April 1998 to June 2000. In fact, prices in PJM were more volatile than prices in the California wholesale market during this period. After June 2000, however, the magnitude and duration of the price spikes are much larger in California.

Electricity price spikes occur as demand approaches the capacity limits of the power system, either at generation sources or at critical transmission points. Sharp, temporary price increases also occur in other industries, such as petroleum refining, natural gas, and many markets for metal commodities. In these markets, consumers often draw from stocks or use financial instruments as insurance for these price shocks. The nonstorability of electricity and the unfortunate configuration of restructuring in California, however, created serious financial problems.

Distribution companies in California, with their retail prices fixed by law, were losing nearly ten cents for every kilowatt-hour sold during the latter half of 2000, and incurring huge financial losses. The problem was further exacerbated by the requirement that distribution companies buy their power on the California Power Exchange spot market, making them unable to shield themselves against price risk through the use of futures and option contracts. By January 2001, the major distribution companies in California were essentially bankrupt. Power generators refused to sell these companies power for fear of nonpayment, and widespread blackouts resulted. The state of California stepped in, eliminated the California Power Exchange, and subsidized electricity markets, at a cost of approximately $40 million per day in the first half of 2001.

Restructuring did not cause the power supply shortage in California. But the form of restructuring—with generators and distributors essentially required to buy on the spot market—exposed the distributors to the risk of relying on spot transactions. The regulated retail prices meant that distribution companies held all the risk. When prices exploded, bankruptcy and blackouts were the natural response. The state of California, by not allowing prices to rise, exacerbated the problem.

The Pennsylvania Experience

The Pennsylvania restructuring plan was similar to the California plan in several ways. Generation was freed from rate-of-return regulation, and power was sold in a largely unregulated market. Generation divestitures were not required, though many took place voluntarily. Prices to consumers were lowered, and capped for the period of stranded cost recovery. Again, prices to consumers were set to the sum of transmission, distribution, generation, and CTC charge (see table 2.2).

Table 2.2.   Average Electricity Rates in Pennsylvania, 1999

		Rate in Cents Per Kilowatt-hour	
Component	PECO	GPU	Allegheny
Generation charge	5.75	4.00	3.22
Transmission & distribution	4.57	3.03	3.06
Transition charge	1.82	0.73	0.64
Amount paid per month	12.14	7.76	6.92

There were, however, two important differences from the California structure. First, power could be sold on a spot or long-term basis, whatever the parties thought was in their best interest. Second, consumers choosing a supplier other than their local distribution company were given shopping credits set administratively by the state Public Utility Commission. Shopping credits were set originally above the generation cost component of retail prices, which allowed retailers to enter the market. Electricity retailers did enter the market, selling at one point up to 10 percent of customers. Unfortunately, as market prices rose (and shopping credits remained fixed), retailers were squeezed out of the market.

Wholesale electricity prices in Pennsylvania in the period 1999–2001 rose by about 25 percent. Power in Pennsylvania comes largely from coal-fired generators, with natural gas plants representing only the marginal suppliers. New power plants are being allowed into the system, though the required administrative and regulatory procedures slow this process down.

The Pennsylvania price cap, just like its California equivalent, does create the possibility of a squeeze on utility margins if wholesale prices rise too high. But that has not happened, and is not likely to. The supply of power in Pennsylvania is very stable, and is not highly dependent on the price of natural gas and on natural factors, such as the amount of rainfall. Summer peaking prices can get very high, but only for relatively short periods of time.

The California Electricity Price Shock

A combination of factors contributed to the spike in wholesale electricity prices in California from June 2000 through early 2001. The chief culprits were an unanticipated surge in electricity demand and a lack of low-cost electricity supply. For more than a decade, it had been extremely difficult to site new power plants in California. The state had become highly dependent on hydroelectric sources, power from natural gas plants, and imported power.

Total electricity sales in California increased more than 6 percent during 2000, exceeding historical average annual demand growth by more than a factor of three. While previous generation capacity could have serviced this load without major disruption, this surge hit just as hydroelectric capacity was about to plummet. The winter of 1999–2000 in the western United States was very dry, significantly reducing the snow pack in Oregon and Washington. In normal years, the summer melt of this snow drives hydroelectric dams, and much of this electricity is transmitted to California to meet its summer power needs. During 2000, however, hydroelectric power generation declined more than 18 percent in Washington and Oregon and 22 percent in the Rocky Mountain region, substantially reducing the availability of power for export to California (see Considine and Kleit 2002). There was also a 13 percent reduction in electricity from other, nonconventional sources of power in California during 2000, apparently associated with difficulties in the production of geothermal power.

The shortage in hydroelectric generation was offset by a substantial increase in natural gas–fired electric power generation. Gas-powered generation increased more than 15,000 million Mwh in California during 2000 and another 14,000 million Mwh in the rest of the western United States. Even under relatively low gas prices, however, the marginal cost of gas-fired electric power is more than three times greater than the marginal cost of hydroelectric power.

This large switch to natural gas placed considerable strain on the natural gas transmission system and substantially reduced gas storage levels. As a result, prices for natural gas delivered to the city-gate in California increased more than fourfold, rising from $2.90 per thousand cubic feet (mcf) in March 2000 to a peak of $12.64 in December 2000 (see figure 2.2). These higher gas prices directly increased the cost of operating natural gas–fired electrical generation capacity.

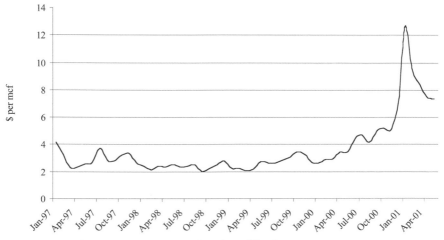

Figure 2.2. City-Gate Natural Gas Prices in California, 1998–2000

The generation cost model developed by Considine (2000) and presented in Considine and Kleit (2002) provides a tool for estimating how these higher natural gas prices affect the marginal cost of electric power generation. Our cost model was estimated using data for utilities before they sold their generation assets. Most of the California utilities sold their assets during 1998 and 1999. As a result, our model only provides estimates of marginal cost for the last year that each utility operated those assets. Nevertheless, the model can be used to estimate what impact higher natural gas prices alone would have on short-run marginal costs of steam electric power generation.

Table 2.3 presents the results from our calculations. We calculated marginal generation cost in the year before divestiture at then prevailing natural gas prices paid by these utilities. Natural gas prices varied between $2.67 and $2.96 per mcf, and the corresponding marginal costs for electricity were between $24 and $34 per Mwh. We then recalculated these prices under average natural gas prices during the year 2000, which were slightly above $6 per mcf. As a result, marginal generation costs rise between $42 and $72 per Mwh (see table 2.3). These short-run marginal costs rise to more than $130 per Mwh at the peak of natural gas prices during January 2001.

A similar story emerges if we take a somewhat different approach to estimate marginal cost. In figure 2.3, we plot the marginal cost of electricity generation for all electricity generators in the state of California during January 2000. Marginal cost is defined here to include the incremental cost, in terms of dollars per Mwh, for fuel, labor, and environmental allowances. We use our cost function estimates for marginal cost from nuclear electric power genera-

Table 2.3. Impact of Higher Natural Gas Prices on Marginal Cost for Electricity

	Southern CA Edison Co	Pacific Gas & Electric Co	San Diego Gas & Electric Co
	1997	1997	1998
Before natural gas price increase			
Gas price ($ / mcf)	2.964	2.713	2.668
percent share of gas in steam generation	96.140	99.10	98.60
Marginal cost ($ / Mwh)	24.561	32.982	33.734
After natural gas price increase			
Average gas price in 2000 ($ / mcf)	6.040	6.040	6.040
Marginal cost ($ / Mwh)	42.670	58.340	72.390
Average gas price in Jan 2001 ($ / mcf)	12.350	12.350	12.350
Marginal cost ($ / Mwh)	75.020	105.760	139.980

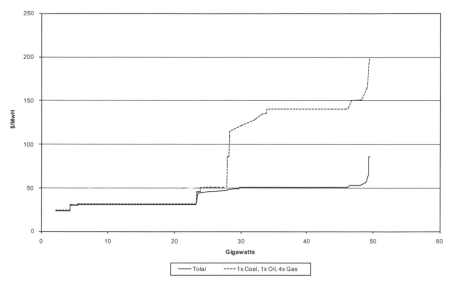

Figure 2.3.   Marginal Cost of Electricity in California, January 2000

tion and estimates of hydroelectric marginal costs used by Considine and
Kleit (2002). The marginal-cost curve is constructed by ranking capacity by
marginal cost. The horizontal segments reflect the relative capacity for each
level of marginal cost. The curve illustrated in figure 2.3 has the characteristic "hockey-stick" shape, with a long, flat handle that includes low-cost
sources of power, such as nuclear and hydro, and a sharp upward section including higher-cost sources, including gas and oil turbines.

This diagram provides a useful reference point to understand the cost implications of the California situation. First, as the diagram illustrates, a fourfold increase in natural gas prices shifts the cost schedule sharply upward at
higher output levels. Second (this is not illustrated in figure 2.3 but easily envisioned), the shortfall in hydroelectric capacity shifts back the entire schedule so that higher cost generators are used earlier in meeting load requirements. If the system was operating at the capacity limit as it no doubt
appeared to be during late 2000, the diagram illustrates that the marginal cost
of the last generating unit would be approaching $200 per Mwh.

This corresponding cost schedule in Pennsylvania, however, is much different. A good example is the corresponding cost schedule for Pennsylvania
illustrated in figure 2.4. Notice that the handle of the "hockey stick" is much
longer, illustrating a greater share of low-marginal-cost nuclear and coal-fired
capacity. This reflects the much greater reliance on coal and nuclear power
generation in Pennsylvania. In contrast, California is far more dependent

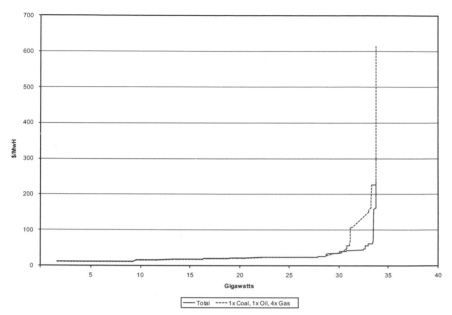

Figure 2.4.   Marginal Cost of Electricity in Pennsylvania, January 2000

upon natural gas and hydroelectric generation, which are far more sensitive to gas prices and hydrological events respectively. Unless demand approaches capacity limits represented by the vertical portion of the schedule, this comparison suggests that price shocks of the magnitude witnessed in California are unlikely in Pennsylvania.

Another implication of this analysis is that there appears to be a rather sizable unexplained difference between wholesale prices in California and the marginal costs estimated in this study under higher natural gas prices during the peak in electricity prices in December 2000. Between June and November 2000, weekly average prices were between $50 and $240 per Mwh, near the boundaries of our estimated cost schedules. Prices during December 2000, however, shot up to more than $600 per Mwh.

There are several possible explanations for this discrepancy. First, our estimates of environmental permit prices may be underestimated and may not reflect the temporary bursts sometimes observed in tightly regulated areas, such as Southern California. Also, transmission capacity constraints may create a congestion premium on electricity prices. Our estimates of marginal cost are at the generation point and do not include such premiums.

A third possibility is that the market price of electricity may have contained a risk premium associated with the financial insolvency of the major utility companies in the state. In this case, generators would place a premium on bids to compensate for any unpaid bills. For example, Pearlson (2001) reported that on January 17, 2001, Duke Power offered to operate an inefficient and polluting generator for $3,880 per Mwh with most of the proposed fee representing a "credit premium," because it estimated that there was only a 20 percent chance that it would ever get paid. The California grid operators reluctantly agreed to pay this high price, figuring that this amount would be far less than the costs of a blackout. The Federal Energy Regulatory Commission (FERC) determined that a "just and reasonable" price would have been $273 and ordered Duke to return the excess. Ironically, there was nothing to return because Duke received only $70.22 per Mwh, precisely the type of result it had feared.

Yet another possibility is that electricity producers were withholding generation capacity at certain critical times to drive market prices upward. Joskow and Kahn (2001) find that there were four times more scheduled and unscheduled plant shutdowns during the fall and winter of 2000–2001 than in the previous year. Other analysts note that these plant maintenance shutdowns were necessary because these facilities were so intensively operated during the previous twelve months. Nevertheless, the closing of these plants at critical times, such as peak power demand periods, would create a shortage and drive prices upward. Other theories of the exercise of market power are more subtle, involving bidding strategies. For instance, generators would offer extremely high bid prices for power on their last units offered for sale, knowing that their competitors would match these bids. This type of signaling strategy works so long as supplies are tight and as long as no competitor seeks to gain advantage by submitting substantially lower prices. Proving whether these strategies were employed and identifying their motivations could be a daunting challenge. As the following section illustrates, however, market forces, may be a powerful disciplinary force against persistent use of such tactics and on the wielding of market power in general.

AFTER THE PRICE SHOCK

After January 2001, the price of electricity in California dropped considerably (see figure 2.5). From January through early May 2001, the price fluctuated between $150 and $250 per Mwh. During May and into June 2001, however, the price dropped to $50 per Mwh and stayed at that level through the end of August 2001. Several factors contributed to this decline. The demand for electricity fell due to a cooler than anticipated summer, weak economic conditions, and conservation efforts, spurred by high prices and perhaps by changes in energy consumption habits due to public appeals for conservation.

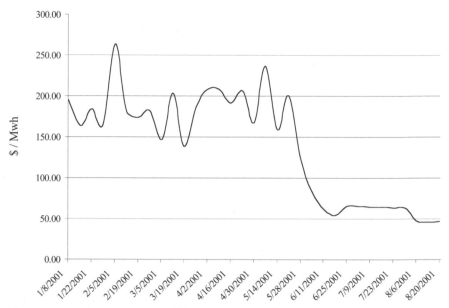

Figure 2.5.   Price on the APX California Energy Market, 2001

Several supply-side factors also possibly contributing to the decline in electricity prices. California elected officials urged that FERC impose tough wholesale price controls during the spring of 2001. During late June, FERC imposed cost-based controls on wholesale electricity prices.

Behr (2001) reports that some analysts have argued that FERC actions were a decisive factor in lowering electricity prices. It appears, however, that market forces were the major factor in the price decline. Natural gas prices became considerably lower, sliding steadily through the spring and summer of 2001 as additional supplies substantially rebuilt natural gas storage levels nationwide. The price for natural gas on the New York Mercantile Exchange steadily declined through the year from roughly $10 per million British thermal units to less than $3 during August 2001.

Another important factor is a very significant increase in electric power generation capability (see table 2.4). During June and July 2001, more than 1,400 MW of capacity came on line in California and more arrived the next two years. This suggests that unless demand increases, the state could be facing a surplus of electric power generation. Indeed, the long-term contracts signed by the state of California to secure future supplies are now under severe criticism as the state has been forced to sell power it acquires at a loss.

All of the new capacity in California is fired by natural gas. A critical long-term concern is whether there will be enough natural gas to supply these

Table 2.4. Generation Capacity by Fuel Type in California and Pennsylvania

Fuel Type	California		Pennsylvania	
	Gigawatts	*Percent*	*Gigawatts*	*Percent*
Coal	0.532	1.1	18.021	53.4
Oil	0.428	0.9	1.963	5.8
Gas	20.989	42.0	2.339	6.9
Hydro	12.226	24.4	0.674	2.0
Alternative	11.522	23.0	1.542	4.6
Nuclear	4.310	8.6	9.200	27.3
Total	50.008		33.740	

plants at economical prices. The evidence thus far indicates that there will be enough gas to supply these plants. The number of wells drilled in the United States increased from approximately 11,000 in 1999 to more than 20,000 in 2001. Given the advances in drilling and resource recovery technology over the past twenty years, these wells are likely to be very productive. The price increases necessary to induce such a response, however, clearly involve significant economic and financial adjustment costs. Given the nature of electricity cost schedules and the volatility of demand due to weather and other random events, adjustment costs may be unavoidable. As the California case illustrates, however, these costs could be reduced with policies that promote more open and flexible markets.

REGULATORY POLICY ISSUES

The restructured electricity industry will face a large variety of regulatory issues. In this section we discuss these issues and the choices public policy makers face.

The Stranded Cost Problem

After World War II there were many proponents of nuclear power. In a famous statement, one advocate asserted that power from nuclear fission would be "too cheap to meter." Nuclear power plants were built across the United States.

Unfortunately, nuclear power construction was subject to serious cost overruns. There are several possible sources for these large costs. First, important environmental concerns motivated extensive regulation of nuclear power plants by the Nuclear Regulatory Commission (NRC). Many critics have suggested that this regulation was excessive and led to larger than necessary costs. Second, many different designs of plants were used, precluding firms from being able to learn from others' mistakes. Third, utilities, which were being reimbursed under rate-of-return regulation, had limited or nonexistent incentives to keep prices down. Finally, the advocates of nuclear power may have simply oversold their product.

In most industries, firms that incur cost overruns suffer financial losses. In a regulated utility industry, however, those losses are simply passed on to consumers. Today, the losses due to nuclear power are generally referred to as "stranded costs."

One important rationale for electricity restructuring is to avoid the recurrence of the stranded cost problem. Removing rate-of-return regulation eliminates any incentive to incur cost overruns, hopefully making the electric power system more efficient in the long run.

MARKET POWER CONCERNS

While the generation of electricity may not be a natural monopoly, this does not imply that a particular electricity market cannot be subject to the exercise of market power, or monopolization. In the United States, almost all industries are subject to antitrust enforcement, referred to in other parts of the world as "competition policy," which seeks to prevent markets from being monopolized. In the electricity area, antitrust review can be undertaken by the U.S. Department of Justice, FERC, and state regulatory commissions.

No market is likely to reach the textbook conditions needed for perfect competition. However, markets are often considered "workably competitive" if there is "enough" competition in them so that they approximate perfectly competitive markets. The antitrust authorities at the Department of Justice and the Federal Trade Commission (FTC) consider markets to be "highly concentrated" if the relevant Herfindahl-Hirschman Index (HHI) is greater than 1800. The HHI is calculated by squaring the market shares of the relevant firms in the market, and then adding the squared terms together.

An analysis of market shares found California electricity markets to be highly concentrated prior to restructuring, and divestitures were accordingly ordered there. In Pennsylvania, markets were not found to be concentrated, and no divestitures were ordered.

Unfortunately, these analyses of markets in California and Pennsylvania did not take account of two factors. First, the demand for electricity is highly variable across time periods. Electricity cannot generally be stored, and this contributes to considerable price variation by season and time of day, depending upon how close demand is to capacity constraints. Second, the analysis failed to consider that all power plants do not have the same cost structures. In particular, there are a number of "peaking" plants that only are economical to operate when the price of power is high.

Consider the incentives of a firm with a great deal of "base-load" capacity, that is, economical to use most of the time, and one plant that is only economical to use when a "peaking" situation occurs. This firm may realize that it can achieve greater profits by not running its peaking facility when electricity prices are high, even when the price of power makes running that plant on a stand-alone basis remunerative. The reason for this is that not operating the peaking plant may increase the market price of power. Thus, by not running the peaking facility, the firm can increase the price that its base-load plants receive, increasing its profits.

The province of Alberta in its ongoing restructuring program has implemented one possible solution to this problem. In this program, the rights to the electricity produced by peaking units cannot be directly acquired by those parties who also own substantial base-load capacity. Because market power problems are apparently generated by this ownership combination, it is hoped that this ownership restriction can result in more competitive markets (see Daniel, Doucet, and Plourde, chapter 5).

The evidence of the exercise of market power in California electricity markets is mixed. The working paper by Harvey and Hogan (2001) casts doubt on the possibility of market power. On the other hand, the working paper by Puller (2001) indicates that electricity producers were exercising market power in California prior to the year 2000. Another working paper by Mansur (2001) argues that suppliers in the PJM market exercise market power. Detecting and measuring market power, however, is quite difficult because it involves measuring marginal cost at the plant or in some cases the generator level for very small intervals of time. Often the data for such estimations are unavailable. We expect research on this issue to continue.

SYSTEM OPERATION

Electricity markets require a system operator who can insure that the delivery of electricity remains reliable and that the supply of electricity equals the demand. In the regulated era, most electric utilities performed their own system operation. However, in a deregulated environment, the suppliers of power are independently owned and operated. This generates the need for an Independent System Operator (ISO, or now Regional Transmission Organizations, or RTOs) to manage the needs of the electricity system.

The difficulty arises because the owners of the transmission system that independent producers wish to use are often competitors of those producers.

Thus, the transmission owners can have incentives to discriminate against independent producers, reducing the efficiency in the market. ISOs act to take over control of the transmission system, hopefully inducing more efficient operation of the system.

Currently there are operating ISOs in California, the Midwest, New England, New York, the Pennsylvania-New Jersey-Maryland area (PJM, now expanding into Virginia, Ohio, and Illinois), and Texas. These ISOs are all organized as not-for-profit entities. FERC has proposed the creation of similar organizations called Regional Transmission Organizations.

In addition to balancing the supply and demand of power, the ISO provides ancillary services such as reactive power (required to keep the system itself operating) and spinning reserve (to deal with sudden power losses). An ISO also acts to plan transmission capacity increases in the relevant region. Each proposed transmission line needs to be evaluated for its contribution to overall system reliability.

Another important role of an ISO is to serve as a platform for trading between parties in the electricity market. When electricity is traded between entities, the transaction is potentially subject to a series of transmission fees, one for every electric utility the contract path of electricity crosses. This "transmission pancaking" can preclude the efficient trading of electricity. By acting as a single entity for the purposes of transmission fees, an ISO can serve to eliminate transmission pancaking across the relevant area. The larger the area of the ISO, the more transmission fees can be reduced, and the larger the gains from trade, all other things being equal.

Several issues arise concerning ISOs. The first is their governance structure. Firms in private industry are run primarily with the purpose of making profits. ISOs, being nonprofit, do not have such an incentive. Presumably, the goal of an ISO is to run an electricity system that maximizes the wealth available to the relevant portion of society. However, defining what policies will achieve those goals is often quite unclear.

ISOs are generally run or are subject to the influence of a group of "stakeholders." These involve representatives from both producers and consumers of electricity. Stakeholders may also include representatives of "public interest" organizations or of local governments. These stakeholders may have interests that differ from those of the public at large.

For example, assume that transmission congestion has led to a proposal for new transmission capacity in a particular area. While such capacity might well be in the public interest, the owner of the current capacity would suffer reduced profits were that new capacity to be installed. The incumbent owner might use the stakeholder process to delay or prevent the transmission capacity expansion, arguing a variety of technical points.

The complete role of the ISO is also not without controversy. ISOs run a series of markets in current power, power to be delivered in the future, and in transmission rights. There is no particular reason to believe that the ISO is the most efficient party to run such markets. It may be the case that markets would perform more efficiently if private, for-profit parties were to run them. On the other hand, perhaps markets need to be oriented in a manner consistent with smoother system operations.

FERC has had a policy for several years of encouraging utilities to form ISOs. Many utilities, however, have not been eager to do so. Thus, the goal of creating ISOs may be difficult to accomplish. An ISO is a complex organization, and if utilities do not wish to form them, they can use such complexity to create a host of administration problems.

TRANSMISSION

Electricity must be transmitted from its place of generation to its place of consumption. In the regulated era, the vertically integrated electric utility was responsible for transmission, and received a rate of return approved by the relevant regulatory commission for its investment in transmission.

Electricity systems have always been connected to each other for reliability reasons. For example, if one system has problems with its operating equipment, it can draw on another system's power through use of transmission lines. Starting in the 1960s, and perhaps earlier, electricity firms began using their transmission networks for trading between systems, as Kleit and Michaels (1994) discuss. For example, systems with summer peaks are often adjacent to systems with winter peaks. Transmission lines allow trade between these two systems, reducing the costs to both systems of supplying power to their customers.

Economic trading grew in the 1980s, as more firms sought the gains from doing so. Trades were discouraged by two factors. First, electric utilities were not required to allow firms to use their transmissions systems to undertake trades. Second, because trades often had to go through many different utility systems, the cost of such trades was subject to pancaking. These several charges served to reduce the opportunities for efficient trade.

The Energy Policy Act of 1992 and FERC Order 888 require that utility companies open up their transmission lines to third-party trades. Pancaking has been reduced by two factors. First, there have been a series of mergers in this industry, reducing the number of firm boundaries any electricity trade must take. Second, the establishment of ISOs has served to create trading platforms where electricity trades do not have to cross any firm boundaries.

In the short run, transmission facilities face the challenge of being priced in order to reflect their scarcity value. During most hours, transmission capacity is not scarce and, therefore, has no opportunity cost. During peak hours, however, transmission capacity can be very scarce, and pricing it is important so that power goes to where it has the highest opportunity costs. The scarcity of transmission lines is governed by physical laws that are not always intuitive. Further, transmission lines are subject to "loop flow," where power sent between two different sources can be transmitted across a third party's lines without compensation going to that third party.

Many of the short-run problems accruing to the use of transmission lines appear to be largely solved in the PJM system. There, each of several dozen nodes has (at least during certain periods of time) its own price of power, and, implicitly, its own transmission costs. Similar transmission pricing systems are used elsewhere.

It is less clear how longer-run problems will be dealt with. Many believe that more transmission lines will be needed in the future. Obtaining permission to build new transmission lines, however, is quite difficult. It often takes approval from numerous political entities, which do not wish to have unsightly (and allegedly unhealthy) power lines running across their jurisdictions. To deal with this issue, the Bush administration in the spring of 2001 proposed allowing builders of transmission lines to use the eminent domain authority of the federal government to obtain permission to build new power lines. Many parties, however, objected to this proposal based on concerns about the environment, and the proposal was not enacted.

The second long-term transmission capacity issue revolves around the incentives for firms to build new capacity. Unfortunately, there seem to be sufficient economies of scale in transmission capacity such that building new capacity, at least in some circumstances, would eliminate (perhaps completely) any transmission scarcity. This would eliminate any return to building such lines. Further, much of the opportunity to build new transmission lines is possessed by current owners, as they are often in a position to expand their capacity.

METERING

One of the advantages of restructuring an electricity system is that it allows for "time-of-day" pricing. Under time-of-day pricing, final customers pay a price for electricity based on the wholesale price of power at the time of usage. Because the price of power can vary widely, and is often much higher in the daytime than at night, there can be important efficiency gains available through time-of-day pricing. For example, an electricity-intensive factory could shift its production from expensive daytime operation to less expensive nighttime operation.

Time-of-day pricing requires sophisticated, modern electricity meters that may be expensive to acquire. Despite this, several commentators have urged that state PUCs intervene to encourage time-of-day pricing (see Doucet and Kleit, 2002). Doucet and Kleit find, however, that in a well-operating wholesale market, there is no economic argument for government intervention in this area. On the other hand, if the wholesale market has market power problems, or is subject to often binding rate caps, the government may wish to encourage metering.

CONCLUSIONS

The regulated electricity system has a number of important structural deficits. Most importantly, it provides poor incentives for cost-reduction and innovation, as well as acting to reduce choices that are available to consumers. By putting the production and marketing of electricity into a competitive market, restructuring offers the opportunity for substantial gains for society. In particular, it has the potential to eliminate the hidden costs of regulation.

This is not to say, however, that electricity restructuring is a panacea. We suggest that real price reductions and increases in consumer choice will occur, but that they may not take place immediately upon the beginning of restructuring. Further, any new efforts at restructuring must take into account the results of previous restructuring efforts. In this light, we have several recommendations for future restructuring efforts.

First, restructuring should not place any limits on trading in wholesale markets. The California Power Exchange is an idea that is an obvious failure. Financial markets in commodities, which the California power exchange precluded, exist for a great number of reasons. In particular, financial markets, using instruments such as futures and options contracts, act to allocate risk efficiently among parties. There is no reason that consumers in electricity markets should be denied access to the market for risk, which exists in many other areas.

There are substantial drawbacks to the use of price caps in both wholesale and retail markets. Retail price caps send the wrong signal to consumers. Electricity became scarce in California in the period 2000–2001. Consumers in California, however, had no incentive to reduce their consumption of electricity, as their prices were fixed. Fixed consumer prices served to both bankrupt utilities in the state and to place tremendous strain on the electrical grid. Variable prices, on the other hand, act to give consumers incentives to seek out the best power supply contract for themselves, as well as to reduce demand during periods of high prices.

There is some reason to support wholesale price caps equal to slightly above the level of costs at the highest generator in a relevant market. The difficulty arises because in markets without substantial metering there may be no finite ceiling to the free market price. We are concerned, however, that wholesale price caps, however temporary they are designed to be, may become permanent. Permanent price caps that are binding for a significant amount of time deter the creation of new electricity supply and may cause power blackouts.

While there was previously a debate over whether stranded cost recovery was appropriate, that debate is now over. An integral part of any restructuring plan is the recovery of utilities' stranded costs. Unfortunately, the method chosen for doing so in California and Pennsylvania was inappropriate. Perhaps, a stranded cost "tax" could be attached to every kilowatt-hour of power sold in relevant areas. Alternatively, households could be assessed a fixed "access fee" that would not vary with usage. Consumers would then be free to contract for their power supply. This would eliminate the need for administratively set "shopping credits," as in Pennsylvania, which should allow market conditions in the retail market to more readily react to changes in the wholesale market.

Innovation in the retail market for power has been limited. The method chosen for the payment of stranded costs has served to deter competitive activity in this area. In Pennsylvania, for example, retailers have been forced to leave the market as the price of power rose above the administratively set shopping credit. Hence, another reason to eliminate such shopping credits is the positive impact that this might have on retail market activity and innovation. Part of the rationale for restructuring is to allow new firms into the market for generation of electricity. This requires that environmental and zoning restrictions on generation allow for the construction of new power plants within a relatively short amount of time. Much of the power problems in California were aggravated by the lack of new power plant construction in the previous decade. In Pennsylvania, by way of contrast, new power plant construction is occurring, though regulatory delays occur. We recommend that any jurisdiction considering restructuring consider carefully how high regulatory barriers to new construction are.

Independent System Operators (or similar Regional Transmission Organizations) serve to facilitate trade between parties in wholesale power markets. As such, they are helpful to reaching the full potential of restructuring. We therefore applaud FERC's efforts to create such organizations. We would also encourage state regulatory agencies to do what they can in this regard. It is important to note, however, that regulators may have difficulty inducing unwilling utilities, which may not be supportive of restructuring efforts, to effectively join in creating ISOs.

Any gains from restructuring can be diminished by the exercise of market power. Jurisdictions having already embarked on restructuring were aware of this problem. They conducted standard market power reviews, and, in the case of California, required divestiture. A new difficulty, however, has arisen due to the structure of the electricity supply curve. During periods of high prices, decisions made by producers of high-cost facilities (which only come on line during these periods) can materially affect the market price of power. If a producer owns both a low-cost facility and a high-cost facility, it may be able to increase its net profits by not operating the high-cost facility, even though the market price may be greater than the cost of operating that high-cost facility. This action would increase market price and harm consumers. We therefore suggest that regulators review carefully the ownership structure of generation in the industry and require the appropriate divestitures.

REFERENCES

Behr, Peter. (2001) "Region's Electric Rates Kept in Check—For Now." *Washington Post*, February 12, p. A1.

Center for Resource Solutions. (2001) Renewable Energy Certification Program. http://www.green-e.org/what_is/what_is_index.html.

Chao, Hung-Po, and Stephen Peck. (1996) "A Market Mechanism for Electric Power Transmission." *Journal of Regulatory Economics* 10 (1): 25–59.

Considine, Timothy J. (2000) "Cost Structures for Fossil Fuel–Fired Electric Power Generation." *The Energy Journal* 21 (2): 83–104.

Considine, Timothy J., and Andrew N. Kleit. (2002) *Comparing Electricity Deregulation in California and Pennsylvania: Implications for the Appalachian Region.* Final Report to Appalachian Regional Commission, ARC Contract Number CO-12884.

Daniel, Terry, Joseph Doucet, and André Plourde. "Electricity Industry Restructuring: The Alberta Experience." Chapter 5, this volume.

Dismukes, David, and A. N. Kleit. (1999) "Cogeneration and Electric Power Industry Restructuring." *Resource and Energy Economics* 21 (1): 153–66.

Doucet, Joseph, and A. N. Kleit. (2002) "Metering in Electricity Markets: When Is More Better?" In *Markets, Pricing, and Deregulation of Utilities*, ed. Michael A. Crew and Joseph C. Schuh, 87–108. Boston: Kluwer Academic Publishers.

Dubin, Jeffrey, and Geoffrey Rothwell. (1990) "Subsidy to Nuclear Power through Price-Anderson Liability Limit." *Contemporary Policy Issues* 8 (3): 73–79.

Energy Information Administration. (1996) *Financial Statistics of Major U.S. Investor-Owned Electric Utilities.* Washington, D.C.: U.S. Government Printing Office.

Ethier, Robert, et al. (2000) "A Comparison of Hypothetical Phone and Mail Contingent Valuation Responses for Green-Pricing Electricity Programs." *Land Economics* 76 (1): 54–76.

Federal Energy Regulatory Commission. (2001) "Mergers and Other Corporate Applications Information." http://www.ferc.gov/electric/mergers/mrgrpag.htm.

Green, Richard. (1999) "The Electricity Contract Market in England and Wales." *Journal of Industrial Economics* 47 (1): 107–24.

Harvey, Scott, and William Hogan. (2001) "Further Analysis of the Exercise of Market Power in the California Electricity Market." Working Paper, Harvard Electricity Policy Group, November 21, 2001. http://www.ksg.harvard.edu/hepg/Papers/Hogan%20Harvey%20CA%20Market%20Power%2012-28-01.pdf.

Heyes, Anthony, and Catherine Liston-Heyes. (1998) "Subsidy to Nuclear Power through Price-Anderson Liability Limit: Comment." *Contemporary Policy Issues* 8, (1): 122–24.

Joskow, P. L. (1996) "Restructuring, Competition, and Regulatory Reform in the U.S. Electricity Sector." *Journal of Economic Perspectives* 11 (3): 119–38.

Joskow, P. L., and Edward Kahn. (2001). "Identifying the Exercise of Market Power: Refining the Estimates." http://web.mit.edu/pjoskow/www/.

Kleit, A. N., and R. T. Michaels. (1994) "Antitrust, Rent-Seeking, and Regulation: The Continuing Errors of *Otter Tail.*" *Antitrust Bulletin* 39 (3): 689–725.

Mansur, Erin T. (2001) "Pricing Behavior in the Initial Summer of the Restructured PJM Wholesale Electricity Market." University of California Energy Institute, PWP 083, April 2001.

Newbery, David M. (1999) *Privatization, Restructuring, and Regulation of Network Industries.* Cambridge, MA.: MIT Press.

Nuclear Energy Institute WebPages. http://www.nei.org/.

Office of the Consumer Advocate. (2001) Commonwealth of Pennsylvania. http://www.oca.state.pa.us/.

Pearlson, Steven. (2001) "The $3,880 Megawatt-Hour: How Supply, Demand, and Maybe 'Market Power' Inflated a $273 Commodity." *Washington Post*, August 20. http://www.washingtonpost.com.

Puller, Steven L. (2001) "Pricing and Firm Conduct in California's Deregulated Electricity Market." University of California Energy Institute, PWP 080, January.

Reiffen, David, and A. N. Kleit. (1990) "Terminal Railroad Revisited: Foreclosure of an Essential Facility or Simple Horizontal Monopoly?" *Journal of Law and Economics* 33 (2): 419–38.

Wilson, Wesley W. (1997) "Cost Savings and Productivity in the Railroad Industry." *Journal of Regulatory Economics* 11 (1): 21–40.

Winston, Clifford. (1998) "U.S. Industry Adjustments to Economic Deregulation." *Journal of Economic Perspectives* 12 (1): 89–110.

NOTE

1. We thank Tim Brennan and Karen Palmer of Resources for the Future; Terry Boston, Wayne Gildroy, and Mark Medford of the Tennessee Valley Authority; Howard Gruenspecht, Doug Hale, Gary Merinholz, and Tracy Terry of the Energy Information Administration; John Hangar of Citizens for Pennsylvania's Future; Udi Helman, David Hunger, Paul Sotkeiwicz, and Roland Wentworth of the Federal Energy Regulatory Commission; John Kelly and Michael Nolan of the American Public Power Association; Steven Kirchoff of the National Rural Electric Cooperative Association; Irwin Popowsky, Consumer Advocate for the State of Pennsylvania; and Samantha Slater of the Electric Power Supply Association for helpful discussions. We also thank Roy Hampton, Joseph Rix, Supawat Rangsuriyawboon, and Bo Yang for excellent research assistance, and Gregory Bischack of the Appalachian Regional Commission for excellent comments. Financial support for this project came through Contract Number CO-12884 from the Appalachian Regional Commission.

3

The Role of Retail Pricing
in Electricity Restructuring[1]

L. Lynne Kiesling

Since the inception of state regulation in 1907, retail customers of electricity have faced average rates that change infrequently, insulating customers from both price increases and decreases. Retail electric service has been provided on a guaranteed-price basis, under the regulatory "obligation to serve" facing regulated utilities. The must-serve obligation on the regulated utility is a relic of the political dynamic of regulated electric power in the United States, in which supply reliability and stable price are the highest priorities, regardless of the cost of providing that reliability. Thus, supply reliability has historically been considered only a supply-side problem, and the industry, regulators, and political institutions have internalized that belief.

Fixed retail rates mean that the prices individual consumers pay bear little or no relation to the marginal cost of providing power in any given hour. Moreover, since retail prices do not fluctuate, consumers are given no incentive to change their consumption as the marginal cost of producing electricity changes. The consequence of this disconnect goes beyond inefficient energy consumption and also causes inappropriate investment in generation and transmission capacity.

The negative consequences of fixing retail rates were hidden for decades by other aspects of regulation, such as the control of wholesale prices and excess supply in generation, but these have become more obvious in the era of restructuring. In particular, the freeing up of wholesale prices disconnected the wholesale and retail markets, with unintended effects. As we should have learned from the savings and loan banking crisis (FDIC 1997), partial deregulation can be worse than no regulation at all.

Without retail pricing reform, electricity restructuring will fail to deliver efficiency and value to consumers. The "one size fits all" of regulated and fixed rates is becoming increasingly obsolete because of technological, institutional, regulatory, and cultural changes that recognize the diversity of products that the electricity industry can profitably sell to consumers. Retail reform is necessary to maximize the value of other market reforms and is in and of itself a valuable step in producing efficient and fair electricity markets.

RETAIL PRICE REGULATION

Background

Regulation played a significant role in the electricity industry almost from the advent of commercial electricity sales in the late nineteenth century. The increasing economies of scale in the generation and distribution of electricity led to the early application of the theory of natural monopoly. Economists say that a firm is a natural monopoly if it can produce the entire market output at a cost that is lower than if the output were produced by several firms (Viscusi et al.). Although a natural monopolist is able to produce at lowest cost, it will not necessarily sell to consumers at lowest price. The unpleasant choice between monopoly pricing and inefficient production or potentially unsustainable competitive pricing at high cost provides a justification for regulation. Electric utilities have typically been regulated under the natural monopoly doctrine using either price regulation or rate-of-return regulation. Under rate-of-return regulation, as implemented in the United States, utilities submit their operating and capital cost data and forecasts to a regulatory agency, which sets a price (or rate) that allows the utility to earn a "reasonable" rate of return on its capital investment. Theoretically, regulation can lead to the best of all worlds—production by the low-cost natural monopoly without monopoly pricing.

The "public interest" theory of regulation mirrors the natural monopoly theory; it says that what should be done in theory is done in practice. In the case of electricity, concern about the monopoly power created by economies of scale in electricity production led the states to introduce rate and service-area regulation under public utilities commissions. But the competing "rent-seeking" theory of regulation argues that state utility regulation arose out of the interests of incumbents in protecting their industry from competition, not from a public or consumer-interest concern about monopoly power and possible price increases. In this view "natural" monopolies turn out be rather unnatural. Jarrell (1978), for example, reported that between 1900 and 1920 the states that initially adopted utility regulation actually had lower prices and profits than the states that did not adopt utility regulation. Jarrell's findings suggest that regulation was adopted first where prices were most, not least, competitive, which is more consistent with the rent-seeking theory of regulation.

Part of the regulation of electric utilities was the granting of exclusive franchises for specific service territories to the utilities (Energy Information Administration 2000). Although the details vary by state, the franchise also generally carries with it an obligation to serve all present and future customers in the service territory at a reasonable cost (Edison Electric Institute 1991). This obligation to serve all customers in a territory persists to this day as a fundamental characteristic of the monopoly franchise and has served to eliminate possible competition for utilities, including competition from new technologies for distributed generation. Basing the rates that customers pay on cost recovery is one of the consequences of the obligation to serve (in combination with rate-of-return regulation). This focus on cost recovery in rates often provides an obstacle to the evolution of market-based retail electric pricing, because instead of considering the value created for customers it emphasizes solely the cost of providing customers with a particular type and level of service.

Even regulators have long acknowledged that the regulated world is at best a second-best environment. The regulated system does not capture the rich and complex web of information that would lead to "efficient" pricing and supply of electricity. In reality, determining capital costs and implementing rate-of-return regulation is complicated, and does not necessarily generate an efficient outcome. Electric utilities provide multiple products and services, so the simple "decreasing long-run average cost/economies of scale" designation as the basis for electricity regulation is likely to lead to inefficient outcomes. Furthermore, regulator decisions regarding what to include or exclude in the allowable costs for determining rates are subject to manipulation.

Retail Prices and Deregulation

Retail customers still overwhelmingly face guaranteed average prices in the form of rate caps, even in states that have implemented some measure of deregulation. The way that institutions have evolved in electricity restructuring in the states, and the political compromises struck to move restructuring forward, have led to an environment in which wholesale electricity markets and retail rates are still largely disconnected. The states that have not implemented any restructuring do not have wholesale price signals, in addition to retaining fixed retail rates. These price caps reduce the incentives of customers to participate in voluntary programs that harness the price elasticity of demand to send informative price signals into the wholesale market. The artificial environment of fixed retail prices largely removes the price signals that would enable consumers to respond and act on those price sensitivities. Fixed retail rates also ignore the diversity among customers in their demand,

and exploiting that diversity would lead to better information in the market, lower costs, and well-served customers.

States in the United States that have pursued electricity restructuring have retained regulated retail rates in a phase-out transition period, as a result of the political compromise to get restructuring passed; "consumer advocates" view regulated retail rates as a consumer protection, largely based on risk aversion. These transition rates are usually at a discount relative to some historic rate; the pioneer in this form of transitioning away from regulated retail rates was the United Kingdom. In Pennsylvania, for example, the regulated retail rate tariff will transition out over ten years, insulating retail customers from price changes for a decade.

THE BENEFITS OF MARKET-BASED PRICING

George and Faruqui (2002, 2) define market-based pricing as "any electricity tariff that recognizes the inherent uncertainty in supply costs." Market-based pricing can include time-of-use (TOU) rates, which are different prices in blocks over a day, based on expected wholesale prices; or real-time pricing (RTP), in which actual market prices are transmitted to consumers, generally in increments of less than an hour. As currently implemented, TOU is typically a program in which predetermined prices apply to specific time periods by day and by season. RTP differs from TOU mainly because RTP exposes consumers to unexpected variations (positive and negative) due to demand conditions, weather, and other factors. In a sense, fixed retail rates and RTP are the endpoints of a continuum of how much price variability the consumer sees, and different types of TOU systems are points on that continuum (Hirst 2001a, 2). Thus RTP is but one example of market-based pricing. Both RTP and TOU provide better price signals to customers than current regulated average prices do. They also enable companies to sell, and customers to purchase, electric power service as a differentiated product depending on the time of day of its use.

The list of benefits from implementing market-based pricing is extensive and widely agreed upon by most analysts. Focusing on market-based pricing emphasizes the information content of prices, an aspect of prices that frequently gets overlooked in political debates. The most important features of market-based pricing and demand response arise when consumers can choose how much of the real cost of power they see over which time period, and when they have the corresponding choice of prices to face (what might be called the "portfolio of contracts concept"). An important policy distinction arises between customers being *required* to see hourly prices, and customers having the *opportunity* to see hourly prices (Hirst 2001b). Requiring real-time pricing would both contradict the idea of choice and expose some customers to more price risk than they would choose voluntarily.

The benefits of market-based pricing fall into three categories: economic efficiency, reliability, and environmental quality. Dynamic retail pricing enables customers to shift demand away from peak periods with high prices, and/or to reduce their overall use. This economizing incentive is the source of the conservation benefits of market-based pricing. Conservation brought on by market-based pricing reduces energy costs and increases energy efficiency (Taylor and Schwarz, 1990). Conservation typically takes two forms—curtailing consumption (reducing overall use) and shifting use to nonpeak hours. The first-order effects of these consumption changes are felt directly by the consumers who choose to curtail or shift use, and about 20 percent of the value of market-based pricing comes from this direct effect (McKinsey & Company 2001, 5). Most of the value of market-based pricing, though, comes through an indirect effect—these responses among individual consumers create a reduction in peak demand, which reduces wholesale prices for all other consumers of all power in that hour. Thus market-based pricing gives customers incentives to manage their own energy costs and helps bring aggregate supply and aggregate demand into balance at levels that do not exhibit prices as high or as volatile as in the absence of demand response.

Critics of this portfolio of contracts vision worry that customer bills would increase and would become unpredictable if customers faced such choices. This concern is overstated. If retail providers were free to choose contractual forms that their customers might want, then consumers are unlikely to desire, and thus to face, higher and unstable bills. If consumers were concerned about price volatility, they could choose contracts with some insurance against price risk; in other words, customers would choose how predictable their monthly bills would be. They may face higher rates in some hours if they choose to do so, but such a choice is likely to reduce consumption in those peak hours with higher rates. The effects of shifting demand away from peak would reduce their use in those hours, and the overall effect on prices in *all* hours could lead to lower electricity bills for consumers.

Critics who focus solely on price, though, miss one of the most important points about market-based pricing and making a portfolio of choices available to consumers—market-based pricing and choice do not guarantee lower average retail rates to consumers, particularly in circumstances where input costs are rising. Market-based pricing and retail choice are likely to drive prices lower in the long run due to more efficient resource allocation, relative to the performance of other institutions under the same conditions. But overlooking the potential that retail choice has of offering consumers services, choices, and opportunities that benefit them much more than the standard electricity service they currently receive means overlooking one of the most important sources of value to customers. Consumers do not always make consumption and contractual decisions based solely on price levels, and focusing solely on prices is unnecessarily myopic.

Market-based pricing also increases competitiveness of electricity markets and reduces the severity of price spikes. Customers modifying their use when they see price volatility help reduce the magnitude of price spikes. When consumers can receive price signals and can respond to them, some consumers, when they face high prices, will shift their demand to cheaper hours. Shifting demand from an expensive hour to a cheaper hour lowers equilibrium price in the expensive hour and may increase it in the cheaper hour.[2] "Marginal cost real-time pricing also opens the door to conservation and load management to all customer classes. Customers will 'discover' avenues to manage load that make economic sense. In most cases conservation and load response will be implemented before taking on the larger capital investments for on-site generation" (Winters 2001, 77).

Another benefit of market-based pricing related to reduced volatility is that it gives consumers a direct way to hold generators accountable for their wholesale pricing decisions. Market-based pricing integrates wholesale and retail markets, by transmitting information that drives retail prices to reflect costs more accurately. That integration means that customers bear wholesale electricity prices more directly, and therefore will be more likely to shift demand away from hours with high wholesale prices. During peak demand periods market manipulators can increase profit by holding supplies off the market and driving prices proportionately higher. If demand were to decrease during price spikes, however, the opportunity for manipulation would decline. Therefore, enabling an active demand side through pricing and choice disciplines firms that could exercise market power in a supply-driven market. In markets that have only one-sided supplier bidding, suppliers are more able to manipulate prices than in double-sided markets, where consumers can express their preferences. Market-based pricing reduces the exercise of market power by changing the shape and size of the peak/shoulder/off-peak demand curves, known as load shaping. Price signals discipline producers and decrease their ability to get higher prices in peak hours. If consumers can and will shift their demand to nonpeak hours, then generators cannot charge the kinds of high prices that they could in California's dysfunctional "market" in 2000 and 2001. Reductions in peak-period consumption also increase reliability of the network.

Rassenti, Smith, and Wilson (2001) tested these market power and demand response hypotheses experimentally, specifically looking at situations in which consumers could choose voluntary service interruption contracts. They compared two bidding systems in a wholesale electricity market—one with only supply-side bidding, and one with both supply-side and demand-side bidding. Their experimental design built in the effect of retail customer demand inelasticity on the wholesale market. The buyer in the wholesale market could be an integrated utility, retail electricity provider or even what is sometimes referred to as a "curtailment service provider." RTP and TOU may be the retail pricing tool by which the wholesale buyer obtains the demand response it needs in order to cut purchases as higher prices. Alternatively, of course, an energy service company may have the ability to cycle air conditioners and water heaters off and on or some other direct device to obtain demand reductions. They also had three different cost levels for generators: base-load (low cost), intermediate cost "load followers," and high-cost peaking units.

Their experimental results indicated that demand-side bidding decreased price volatility, the magnitude of price spikes, and the ability of suppliers to exercise market power. Average prices were lower when both demand-side and supply-side bidding existed, as opposed to the one-sided supply-side bidding that characterizes wholesale electricity markets in the United States.

Rassenti, Smith, and Wilson modeled voluntary demand interruption contracts as the tool through which buyers could respond to price changes (although others do exist, such as RTP and TOU, and are likely to be attractive to some consumers). Such contracts capture the information content of prices, as well as the financial insurance component of uninterruptible power prices.

Another related benefit from market-based pricing is often overlooked—removing the rate averaging that comes with regulation removes the subsidies that off-peak users pay to peak users. With fixed, average rates, customers face no incentive to reduce their consumption in peak hours, even though generating electricity is more expensive in peak hours than in off-peak hours. In essence, then, the rate averaging across time subsidizes peak-hour use of power.

Policymakers often express concerns about the volatility of market-based retail prices, because they would reflect wholesale prices. Wholesale electricity prices are inherently volatile for several reasons (Hirst 2001a). Different types of power plants have different generation costs, from low-cost hydro or nuclear to intermediate-cost coal-fired plants, to high-cost natural gas plants, which lead to price variations. In combination with the difference in generation costs, system loads vary hourly, and particularly in an environment with no demand responsiveness there will be hourly variation in system loads. Third, electricity's lack of storability reduces the time over which the network has to be balanced continuously. A fourth variable is unanticipated supply effects due to outages, either from generation equipment failure, transmission equipment failure, or weather effects. Finally, there is an asymmetry in price increases and decreases because of generator start-up costs—in some hours prices can be zero or even negative because of the costs of stopping and restarting a generator. These costs give generators an incentive to take prices below long-run marginal cost, in order to avoid having to shut down a generator.

However, concerns about retail price volatility are exaggerated, especially in an environment where suppliers are free to offer a portfolio of contracts to their consumers. One of the most valuable benefits of market-based pricing, but also one of the most underappreciated and least understood, is its insurance aspects. Demand-responsive prices can provide two types of insurance: financial and physical. Financial insurance is protection against price volatility; physical insurance is protection against quantity volatility. From this point of view, the current regime has too much price insurance.

The financial insurance benefit of market-based pricing derives from the inherent volatility of electricity prices. The traditional fixed average rate for electricity has two components—the price of the electricity commodity itself, and the risk premium that consumers pay for being protected from volatile prices, because consumers can buy as much electricity as they want at a rate that has been set well in advance of their consumption (Hirst 2002a, 1). However, given that regulated rates are typically set to approximate long-run average cost, consumers do not always pay a full insurance premium for the extent that they are insured against price volatility.[3] Furthermore, in states that have pursued restructuring, the political bargain usually includes a fixed, discounted retail rate during a multiyear phase-out of price caps. Discounts on historic rates can exacerbate the extent to which consumers do not pay a full insurance premium for the protection from volatility that they enjoy.

Market-based prices would provide the opportunity for consumers to choose how much of that price risk they are willing to bear, and how much they are willing to pay to avoid it. Although regulated rates have provided financial insurance, they do not fully communicate the cost of insuring different types of consumers against different types of price risks, and to the different degrees that diverse consumers would choose to be insured. Diversity of consumers means that they have, among other things, different risk preferences, and different willingness to pay to avoid price risk. Market-based prices allow the electricity commodity price and the financial insurance premium components of the price to be unbundled, and to be offered separately to customers. This unbundling would enable more efficient pricing of the financial risk, leading to optimal risk allocation.

Suppose, for example, that electric service were priced like cell phone service: $40 per month, 400 anytime kilowatt hours, and then real-time prices or time-of-use rates for consumption beyond 400 kilowatt hours. Suppose further that consumers could choose from a fixed-price contract, a real-time pricing contract, and several different such packages. Then customers would self-select into the contract that enabled them to bear the price risk that best matches their preferences. Some customers would choose full real-time pricing, bearing the entire commodity price and the risk of its fluctuation, but the ability to choose different packages effectively unbundles the commodity price from the insurance premium and enables customers to choose how much to pay for each.

Quantity volatility, and the associated outage risk, is an issue of reliability of service that is not often connected with the idea of insurance, unless you consider part of your rate payment as payment for 100 percent service all the time. This physical insurance characteristic is what creates the opportunity for value in interruptible contracts, as seen in the Rassenti, Smith, and Wilson research discussed above. Most observers also do not incorporate the fact that market-based pricing enables some customers to shift load to off-peak (a form of physical insurance), which can benefit *all* consumers because it would reduce overall prices. So market-based pricing, even if only chosen by a few customers, can benefit all customers. The consumers who choose to use meters and face real-time market-based pricing will provide their own financial insurance, or not, as they choose. But in so doing they may provide a physical spillover benefit to other consumers, by reducing overall peak usage and improving reliability for all, with less excess capacity, and therefore at lower average cost.

Comparing electricity with collision insurance and financial markets illustrates two important facts (Hirst 2002a). First, consumers have experience in dealing with risk tradeoffs, because they see this relationship in other contexts, like collision insurance. Second, different customers have different risk profiles and different risk preferences, and offering them alternatives that capture those differences improves economic efficiency and resource allocation in the industry. For these reasons, if regulators allow customers to choose how much risk to manage, how much to pay to avoid risk, and how to manage those risks, consumers will themselves create physical insurance for the whole system. Thus market-based pricing creates reliability.

Customer load reduction can also serve long-run reliability functions, by reducing the likelihood of transmission bottlenecks and insufficient generation leading to blackouts (Hirst 2002b; Rassenti, Smith, and Wilson 2001, 2002). In the old vertically integrated industry, generation planning and quantity were the focus of the utility. Reliability in this old model required the utility to have enough generation capacity to satisfy *all* demand at *all* hours of the day—this high capital requirement is one consequence of the regulated "obligation to serve" aspect of the government-granted monopoly franchise.

Historically, the excess capacity to enable utilities to meet all possible demand reliably has been very expensive (and those expenses have spurred deregulation in many states). There is no theoretical reason for the focus on suppliers to provide one very high level of reliability. Market-based pricing, by integrating wholesale and retail markets and by enabling consumer demand management as a reliability tool, therefore leads to lower long-run average cost because of avoided investment.

Environmental benefits (particularly reduced power plant emissions) would be another consequence of market-based pricing, through reduced generation and emissions during peak periods in locations and situations in which the concentration of emissions during peak hours creates air quality problems. The shift away from peak consumption reduces the required investment in generation capacity to satisfy peak period demand. Customers shifting and reducing demand would lead to less investment and new plant construction, which would also reduce long-run average cost by reducing capital costs. Thus market-based pricing could lead to both environmental benefits and reduced production costs, although more research is needed to explore the parameters that would bring about this environmental benefit (Holland and Mansur 2004).

The absence of dynamic retail pricing while states have begun deregulating wholesale markets in the past decade has revealed the importance of integrated wholesale and retail markets. Restructuring will not be complete until there are more complete linkages between wholesale and retail markets. Such linkages would decrease the incidence of short-run supply and demand imbalances, such as the imbalances in California in 2000–2001. Caves, Eakin, and Faruqui (2000) analyzed the electricity price spikes in the Midwest in 1998 and 1999; they argue that these price spikes were caused precisely by the disconnectedness of wholesale and retail markets. Keeping retail prices fixed truncates the information flow between wholesale and retail markets, and leads to inefficiency, price spikes, and price volatility.

The benefits of market-based pricing therefore can include conservation, lower costs, reduced wholesale price spikes, reduced ability to exercise market power, higher reliability, and environmental benefits (Fox-Penner 2001, 1). Furthermore, the benefits of market-based pricing are greatest precisely when the need is greatest—when demand is close to capacity and when prices are more likely to be volatile due to unanticipated effects on demand and/or supply. McKinsey & Company (2002) estimates that if consumers had market-based pricing choices, the benefits would include 250 peaking power plants that would not have to be built; avoided transmission and distribution investments; 31,000 tons of nitrogen oxide (NOx) not emitted; reduced water use, natural gas demand, and natural gas transmission; reduced productivity loss from blackouts; and enough saved electricity to supply seven million homes.

EXAMPLES OF MARKET-BASED PRICING

Several utilities have implemented some limited market-based pricing programs. Although small and exploratory, these have generated positive results that will be useful as more utilities move to market-based pricing. None of these programs implements true market-based pricing, though; instead they are "demand-response" programs that use time-of-day price changes to give customers incentives to shift load. That said, they do indicate how powerful price incentives can be for consumers, and how market-based pricing contributes to a reliable, efficient electricity system. The programs fall into two general categories: historical evidence and recent experiments.

Historical Evidence

Over the past three decades, economists have explored the theoretical effects and empirical feasibility of market-based pricing. Most of these older studies tested peak-load pricing (which was a significant issue in many industries in the 1970s and 1980s) and TOU pricing systems, not RTP as it has come to be understood. Based on Boiteux's (1960) seminal article on peak-load pricing, the literature on peak-load pricing in electricity indicated that time-differentiated rates send customers price signals that convey changes in weather conditions, capacity, and other factors that can increase short-run marginal cost.

Caves and Christensen (1980) and Caves, Christensen, and Herriges (1981) describe a residential TOU pricing experiment in Wisconsin between 1976 and 1980. This rate design had two daily periods, peak and off-peak, and different customers had different "slopes" or differences between off-peak and peak rates. These authors found that when they define different hours or groups of hours as different commodities (e.g., peak and off-peak) and that consumers face different prices for those commodities, consumers respond by shifting their use. Furthermore, the consumers whose behavior changed the most were those with air conditioners and those with electric water heaters. The price elasticity of demand of these consumers was higher in certain peak hours, and varied across the day, as measured by differences in elasticities of substitution. Aigner and Hirshberg (1985) found similar results for small and medium-sized commercial and industrial firms, and Caves, Herriges and Keuster (1989) performed a similar analysis of Pacific Gas & Electric's TOU rate experiment, with similar results.

Herriges et al. (1993) analyzed a TOU and a (revenue neutral) RTP experiment performed with large energy customers of Niagara Mohawk. Their econometric analysis indicated that the RTP rates led to smaller increases in total energy use relative to the TOU rate tariff of a control group. In peak hours the RTP users reduced their consumption by 36 percent, while the control group only reduced their peak use by 5 percent. On the highest priced days, the RTP users decreased their energy use, while the control group's use increased. These results suggest that large users do respond to price signals and can both decrease energy demand and shift energy use to nonpeak hours. Herriges et al. also found that responsiveness did vary, even among large users, but that the responses of a few large customers were sufficient to cut peak demand substantially.

In some states, the regulated retail rates disappear once the utility has recouped the legally agreed upon stranded cost. In San Diego in May 2000, for example, the local utility had recouped its stranded costs and could therefore begin charging rates unregulated by the tariff set by the California Public Utility Commission. San Diego Gas & Electric (SDGE) set its rates to end-use customers based on a five-week moving average of wholesale market prices. Unfortunately, the price of natural gas had risen by then, and much of California's "market" had shifted to the real-time spot market, which raised wholesale prices. San Diego Gas & Electric passed these increased costs on to consumers, and in the summer of 2000 most San Diego customers saw their electric rates double. Consumers complained, and complained enough to have rate regulation reimposed in September 2000, but they also conserved in response to price increases. Bushnell and Mansur (2001) estimated that the average price elasticity of demand during the three months before the reimposition of regulated rates was –0.068. This event provides some evidence that, although demand for electric power is inelastic, it is indeed downward sloping, and customers can and do respond to price signals.

Recent Experiments

A recent Government Accountability Office analysis of demand response provides further useful background on the programs discussed here (GAO 2004).

Georgia Power has implemented a real-time pricing pilot program, in which 1,639 commercial and industrial customers participate (O'Sheasy 2002a). These customers typically have 5,000 megawatts of peak demand, and are thus large users relative to most household and commercial customers. Georgia Power's pilot shows how load-profiling practices can be used effectively. Their basic real-time pricing program gives the customer a right to consume the current load profile used in rate calculations for the customer, with deviations from the load profile priced with reference to a real-time price. Thus the customer can consume along the pattern that the utility expected when calculating the regulated rates, and that consumer would be no worse off. The consumer can also choose to deviate at the real-time price. This program uses load profiling to send the appropriate price signals to the consumer, at least for the energy portion of the bill.[4] Monthly administration fees charged to customers range from $155 to $195 depending on plan and usage, to cover billing, administrative, and communication costs. Customers also have access to an Internet website for the retrieval of price information.

Georgia Power has seen load reductions of 500 to 1,000 megawatts (on the order of 10–20 percent of peak demand for these customers), at price differences ranging from $0.50 to $2.00 per megawatt hour (Braithwait 2002, 7). Georgia Power has also observed that its commercial and industrial customers exhibit a wide range of price elasticities of demand when they can act on their preferences—even within one two-digit Standard Industrial Classification (SIC) code (SIC 20: food products), elasticities range from 0.001 to 0.5 (Braithwait 2002, 10). These customers were able to shift demand away from peak hours, reduce overall demand, and smooth out both the prices and the load profile of Georgia Power's large users. According to O'Sheasy's review of the program, "GPC points to several factors that contribute to their remarkable success, but two of the most powerful success enablers are (1) tariff design, and (2) a family of complementary RTP products" (O'Sheasy 2002a, appendix A).

The New York Independent System Operator (NYISO) also initiated demand-response programs aimed at large industrial and commercial customers in the summer of 2001, to help manage peak load during summer months. NYISO's approach incorporated both price-responsive demand and interruptible or curtailable contracts. Their price-responsive demand program, Day-Ahead Demand Reduction Program (DADRP), enables customers to bid into the day-ahead market for load reductions they would make on the following day. Through this bidding process they discover the price they would be paid to commit to a scheduled load reduction on the following day; once the time period for which they scheduled the reduction is complete, the customer is paid for the reduction. The interruption program, Emergency Demand Reduction Program (EDRP), is a day-of-interruptibility contract that is available for hours when there is a shortfall in reliability reserves. Customers can choose to allow the NYISO to interrupt their service, for which the customer is paid a price determined through a bidding process.

According to the authors of a study analyzing the performance of these demand-response programs, DADRP and EDRP prevented outages in summer 2001, and contributed to price reductions and to stable electricity prices (Neenan et al. 2002). Furthermore, the authors indicate a result consistent with the results from experimental economics—a little demand response goes a long way to creating lower, stable prices. Boisvert, Cappers, and Neenan (2002) also emphasize that one reason for the success of these programs is the integration of the demand-response computer systems with the ISO's pricing, scheduling, and balancing operations.

In another article on the NYISO demand response programs, Neenan, Boisvert, and Cappers (2002) look more closely at the customers who chose to participate in the EDRP demand-response programs in 2001, focusing on the risk trade-offs that these customers were willing to make. For example, during three particularly hot days in August 2001, almost 300 EDRP customers curtailed 420 megawatt hours, on average. During these three days NYISO's demand hit its all-time high, but there were no forced outages.

Using a logic analysis of whether a customer chose to interrupt in a particular hour, given the prices facing them, Neenan, Boisvert, and Cappers calculated implicit price elasticities of demand. The average implicit price elasticity of demand they found was –0.09. During periods when customers could choose to have their service interrupted under this program, EDRP participants chose curtailment of 38 percent of their load. They also found that the most responsive customers, those who were most willing to curtail or shift load, were the larger users. This calculation illustrates two very important features of electricity demand that demand-response programs (and market-based pricing) can exploit. First, most observers commonly believe that the demand for electricity is price inelastic, but that is different from the observation that elasticity changes over the day, as well as across different consumers. These results suggest that the demand for electricity is relatively price inelastic in general, but that consumers are willing to shift load over time. Second, it reinforces the point that selective curtailment and interruption among large users with relatively higher price elasticities of demand reduces and stabilizes prices, as well as maintains system reliability.

The demand-response program at Puget Sound Energy in Washington State differed in many ways from those of Georgia Power and NYISO. Puget Sound Energy (PSE) characterizes its demand-response "Personal Energy Management" program as a conservation program primarily for residential customers. The Personal Energy Management program began in May 2001, and three hundred thousand customers enrolled in the first six months. This program is a typical TOU program, with lower prices in off-peak periods and higher prices in peak hours (PSE uses a four-period TOU framework). PSE also provides automated meter reading services and real-time pricing data to customers. Customers can see their daily electricity use according to the four time periods through PSE's website. Typical summer 2001 rates varied from 4.7 cents per kilowatt hour for overnight hours to 6.25 cents per kilowatt hour in peak morning and evening hours.

PSE found that in the first several months of the program, participating customers shifted approximately 5 percent of their demand away from peak hours; in addition, participating customers reduced their overall electricity use. Almost 90 percent of customers took some action to manage their own electricity use when facing TOU pricing. Customers also expressed satisfaction with the program, with 85 percent saying in a PSE survey that they would recommend the program to a friend (Swofford 2002).

In September 2001 PSE extended its Personal Energy Management Program to twenty thousand of its business customers, who had been monitoring their peak load and off-peak load in anticipation of moving to a TOU rate structure. In late 2001 PSE applied to the Washington Utilities and Transportation Commission to make the TOU rate structure a permanent option in PSE's rate tariff, and the customer base grew to almost three hundred thousand. Since then, however, the program has had some difficulties. When PSE sent customers quarterly reports showing that their monthly bills were slightly higher (approximately 80 cents per month) on average than they would have been under regulated average rates, customers began leaving the program. In November 2002 PSE asked the Washington Utilities and Transportation Commission for permission to end the pilot program ten months early. Analysts generally attribute the failures of PSE's program to poor rate structure design and pricing decisions.

Gulf Power in Florida also operates a residential demand-response program, based on a combination of metering and control technology, customer service, and a TOU pricing structure. Borenstein, Jaske, and Rosenfeld (2002, appendix B), and GAO (2004) provide a thorough background analysis of Gulf Power's Good Cents Select program, which uses a four-part TOU price structure, a programmable thermostat that allows customers to establish settings based on both temperature and price, meter-reading technology, and load-control technology for customers to shift load if they choose in response to price signals. Customers also pay a participation fee, which is one unusual feature of the Gulf Power program.

In 2001, two thousand three hundred residences received service under the Good Cents Select program. In that year Gulf Power achieved energy use reductions of 22 percent during high-price periods and 41 percent during critical (usually weather-related) periods. Furthermore, customer satisfaction was 96 percent, the highest satisfaction rating for any Gulf Power program in its history, notwithstanding the monthly participation fee. Customers say that the $4.53 fee (which covers approximately 60 percent of program costs) was worth the energy management and automation benefits that they derived from participating in the program (Borenstein et al. 2002, appendix B).

Another innovative residential demand-response program is in place in northern Illinois. The Energy-Smart Pricing Plan (ESPP) is a joint effort between the Center for Neighborhood Technology's Community Energy Cooperative and Commonwealth Edison. In its first year, the program had more than seven hundred fifty participants in a variety of neighborhoods and types of homes, from large single-family homes to multiple-unit buildings. Commonwealth Edison provides the hourly prices, on a rate tariff approved by the Illinois Commerce Commission.

The keys to the Energy-Smart Pricing Plan are simplicity and transparency in the transmission of information to residential customers. Participants receive a simple interval meter, and can either call a toll-free phone number or visit a website to see what the hourly prices will be on the following day. Furthermore, if the next day's peak prices are going to exceed 10 cents per kilowatt-hour, customers receive a notification by phone, e-mail or fax. Customers will never pay a price above 50 cents per kilowatt-hour, which the Community Energy Cooperative implemented by buying a financial hedge at 50 cents.

In 2003, the first year of the program, customers saved an average of 19.6 percent on their energy bills (Violette 2004). They generally joined the program expecting to save $10 per month on average, and were not disappointed. Surveys indicate that the participants found the price information timely, and that with this small inducement to save money on their energy bill by making small behavioral modifications, they actually became more aware of their energy use overall, not just in the approximately thirty hours last summer that had higher prices. They also said that their personal contributions toward reduced energy use and improving the environment by participating in this plan really mattered to them.

Although the summer of 2003 was mild in northern Illinois, participants did respond in the few hours that prices rose. Most residents responded by increasing the temperature on their air conditioners or shifting their laundry time to off-peak hours. The econometric analysis of the results showed a price elasticity of demand in those hours, at the margin, of –0.042 percent. In other words, when price rose by 100 percent, participants reduced their electricity use by 4.2 percent. For residential electricity customers, this is a healthy response, particularly given the lack of severe weather conditions. This reduction in use is a reduction at the margin, a margin that can often see prices go up by more than 100 percent in peak hours on hot days. Thus although the elasticity number may sound low, because it is at the margin and at the right time, it can take strain off of the system and contribute to grid stability and service reliability in those hours. On average the residents on ESPP reduced their energy use in high-price hours by approximately 20 percent, a number similar to the reductions seen in the Gulf Power program.

The success of such programs for such a heterogeneous variety of customers shows the potential future for active retail choice in electric power. Current "load profiling" practices of public utilities with flat rates lump all consumers into large groups and charge them similar rates whether they consume on-peak or off. This practice means the more frugal customers end up helping to pay for the most extravagant—a kind of "customer service" that belongs to the past.

OBSTACLES TO MARKET-BASED PRICING

Obstacles to market-based pricing include metering and billing technology, existing state-level regulations, regulatory uncertainty at both the state and federal levels, and seventy-five years of embedded culture that is resistant to change.

Market-based pricing requires infrastructure that communicates the prices clearly to customers, and that enables the supplier to meter, record, and store actual usage by time period that corresponds to the time period over which prices change. Conventional metering only allows average pricing, and does not present the energy consumer with any incentives to reduce peak use, or to shift from peak to off-peak use (peak shaving).

Sophisticated metering, billing, and response technologies are widely available and can be implemented cost-effectively, especially for large energy consumers. All of the technology to implement market-based pricing does exist, but the technology has not evolved sufficiently yet to have settled on an industry standard. This lack of real-time metering has an effect on demand elasticity, and many customers have never developed the habit of thinking about how much their electricity costs and different ways of using it and paying for it. The lack of information that sophisticated metering would create means that linking consumer choices to wholesale prices is impossible. Because neither the distribution utility nor the retail provider (if they are separate) can observe the customer's actual consumption patterns during the day, utilities assign customers a statistical load profile that is an average of similar consumers. These load profiles do not reflect the customer's actual usage, and unless consumers and producers can observe that actual usage, consumers cannot benefit from demand shifting, voluntary interruption, or use curtailment.

Doucet and Kleit (2002) show that the incentives to install meters are not independent of other features of regulation. However, they find that under market price caps, customers do not receive all of the social benefits of sophisticated meters. Price caps reduce the incentives of consumers to adopt meters, as can market power (market power can also increase the incentives for meters inefficiently, depending on the structure of the oligopoly game). Metering incentives are best only when electricity is supplied in a competitive market with no price restrictions. Metering is best encouraged, therefore, as part of a coordinated plan to introduce competitive prices.

The investment incentives facing utilities, which currently own the simple watt-hour meters installed in most buildings, are heavily dependent on the rules that regulators devise for billing and access to meter data. Utilities do not have strong incentives to install sophisticated meters if they either will not own them, or will not reap the profits that could come from using them. In addition to the issue of meter ownership and meter use, the issue arises of who will pay for the meters. Utilities currently hesitate to invest in new meters because of ongoing questions regarding whether the metering function will stay with the retail utility; if not, then the investment could become "stranded." Countries such as England, Wales, and Australia have successfully implemented competitive metering and billing through providers other than the incumbent utility, but utilities in the United States have for the most part resisted opening these functions to competition. Until the regulatory uncertainty surrounding these functions is reduced, utilities are unlikely to install sophisticated meters.

Existing government regulations form another set of obstacles to market-based pricing, both at the state and federal level. The traditional rate structure, fixed by state regulation and slow to change, presents a substantial barrier to demand responsiveness and a two-sided retail market. Utilities typically use predetermined load profiles to estimate each customer's use and calculate the customer's bill. This practice significantly inhibits market-based pricing because it ignores individual customer variation and the information that customers can communicate through choices in response to price signals. Furthermore, the persistence of standard-offer service at a discounted rate stifles any incentive customers might have to pursue other pricing options. In state-level regulation, standard-offer service requires the distribution utility to offer a discounted and average electricity price that does not vary. The discount, though, means that the price does not reflect the financial insurance benefits of fixed electricity prices articulated above, and underprices the financial insurance component of electricity prices.

Other government barriers include regulatory uncertainty about future government regulations and market design, including price caps, market power mitigation, and *ex post* refund power. All of these regulatory tools have been considered in dealing with the fallout from the California experience of 2000–2001, and have contributed to the uncertainty of the institutional environment of the industry. This uncertainty tends to make participants stick with the known, if unsatisfying, regulatory pricing environment. At the federal level, Federal Energy Regulatory Commission (FERC) price caps, whether soft or hard, also reduce customer incentives to participate in voluntary demand-response programs. At the state level, Independent System Operators

(ISOs) can stifle demand response through their focus on wholesale markets, where there are fewer participants that trade larger volumes. But ISOs can still play a role in developing demand responsiveness, by accepting different types and sizes of market participants that are not able to participate in ISO system balancing as it is currently arranged (Hirst 2002c, 10). Finally, the lack of coordination between state and federal jurisdictions, and between wholesale and retail institutions that affect retail demand, have increased the transaction costs of moving to a more demand-responsive, integrated electricity market.

Perhaps the most important, but most intangible, obstacle to market-based pricing is inertia. The primary stakeholders in the industry—the utilities, the regulators, and the customers—all have status quo bias. Incumbent utilities, which in this industry are the players with deep pockets, face incentives to maintain the regulated status quo to the extent possible given the economic, technological, and demographic changes surrounding them (Coyle 2002). Customers also have inertia because they have not had to think about their consumption of electricity and the price they pay for it. Regulators and customers explicitly value the stability and predictability that the vertically integrated, historically supply-oriented and reliability-focused environment has created. But what is unseen and unaccounted for is the opportunity cost of such predictability—the foregone value creation in innovative services, empowerment of customers to manage their own energy use, and use of double-sided markets to enhance market efficiency and network reliability.

The utility, the regulator, and the customer all have decades of experience and tradition underpinning their current decisions, and a disinclination to change. As long as consumers, utilities, politicians, and regulators buy into the argument that electricity prices should not vary over time, consumers will not experience the opportunity to benefit from choosing how to purchase electric service, and how much of the price risk to bear themselves. In many cases, that belief permeates utilities, regulators, and customers. The status-quo bias remains one of the most potent obstacles to customer empowerment and system reliability through retail choice.

CONCLUSION

The electricity industry is undergoing substantial change—regulatory, technological, and economic. To deliver the benefits of electric power service to the increasingly heterogeneous range of customers, business models in the industry must also change. Market-based retail pricing is a crucial component of the ability to deliver choice and value to customers. Fixed, regulated average rates are an obsolete relic of a regulatory approach that, if it persists, will stifle creativity.

Customer response to price changes benefits not only those who respond, but also other customers and the system as a whole. Reducing peak use reduces wholesale market prices and long-run investment requirements that affect all customers, not only those who choose to see price changes. Giving customers the right and ability to say "not at that price" is a powerful tool for both efficiency and fairness.

Utilities can also benefit from demand-response programs. The historic development and regulation of the industry has led to a culture in which the prevailing business model for a utility is "sell more power, make more profit." In this world, utilities are prone to perceive the load reduction that can come from demand response as a direct assault on their profits. But what demand response shows us is that electricity can be sold as a differentiated product according to time, not just as homogeneous electrons. Furthermore, that differentiated product can be priced in ways that reflect the true cost of selling it in that hour.

Demand response opens up the possibility that utilities can make more profit by selling less power. But they have to see it as a viable business proposition, and regulators have to show leadership in enabling utilities to offer their customers a portfolio of contracts from which to choose, even those that include choosing to pay higher prices some of the time.

Customer choice and demand response can also reduce the utility's costs in the long run. The level of peak demand determines investment in generation and transmission assets, and the more extensive demand-response programs become across all types of customers, the longer is the time frame between costly and unpopular investments. Unfortunately, in the current regulatory environment that is based solely on cost recovery and profit as a rate of return on assets, neither the utility nor the regulator has incentives to provide the means for saving on future investment.

If utilities, regulators, and politicians consider the possibility that utilities can offer different value propositions to their customers than just "juice coming through the wall," utilities can benefit from using market-based pricing as a tool for offering an attractive portfolio of service options to their customers. Creating value from this change, though, requires vision, and getting the transitions and the institutions right can be extremely difficult. Consumers will have to change how they think about buying electric service, and what that service is, exactly. For that change to occur, politicians and regulators will have to act on the leadership and vision that would allow consumers to take responsibility for their individual purchasing choices.

REFERENCES

Aigner, D., and J. Hirshberg. (1985) "Commercial/Industrial Customer Response to Time of Use Prices: Some Experimental Results." *Rand Journal of Economics* 16, pp. 341–55.

Boisvert, Richard, Peter Cappers, and Bernie Neenan. (2002) "The Benefits of Consumer Participation in Wholesale Electricity Markets." *Electricity Journal* 3, pp. 41–51.

Boiteux, M. (1960) "Peak Load Pricing." *Journal of Business* 2, pp. 157–79.

Bonbright, J. (1961) *Principles of Public Utility Rates*. New York: Columbia University Press.

Borenstein, Severin. (2002) "The Trouble With Electricity Markets: Understanding California's Restructuring Disaster." *Journal of Economic Perspectives* 16, pp. 191–211.

Borenstein, Severin, Michael Jaske, and Arthur Rosenfeld. (2002) "Dynamic Pricing, Advanced Metering, and Demand Response in Electricity Markets." Center for the Study of Energy Markets Working Paper 105, October. Available at http://repositories.cdlib.org/ucei/csem/CSEMWP-105.

Braithwait, Steven. (2002) "Real-Time Pricing and Demand Response: Five Basic Facts About Power Markets and RTP." Presentation to the California Energy Commission, March.

———. (2003a) "Demand Response Is Important—But Let's Not Oversell (Or Over-Price) It." *Electricity Journal* 16, pp. 52–64.

———. (2003b) "Demand Response: The Power of Pricing." *Metering International* 1. Available at http://www.metering.com/archive/031/23_1.htm.

Braithwait, Steven, and Ahmad Faruqui. (2001) "Fixing the California Pricing Mess." *Electricity Journal* 5, pp. 71–72.

Brennan, Tim. (2002) "Challenges in Deregulating Electricity: Drawing Lessons from the California Experience." Presentation to U.S. Department of Energy, March.

Bushnell, James and Erin T. Mansur. (2001) "The Impact of Retail Rate Deregulation on Electricity Consumption in San Diego." PWP-082 www.ucei.berkeley.edu/PDF/pwp084.pdf, April.

Caves, Douglas, and Laurits Christensen. (1980) "Residential Substitution of Off-peak for Peak Electricity Usage under Time-of-Use Pricing." *Energy Journal* 1, pp. 85–142.

Caves, Douglas, Laurits Christensen, and Joseph Herriges. (1981) "The Neoclassical Model of Consumer Demand with Identically Priced Commodities: An Application to Time-of-Use Electricity Pricing." *Rand Journal of Economics* 18, pp. 564–80.

Caves, Douglas, Kelly Eakin, and Ahmad Faruqui. (2000) "Mitigating Price Spikes in Wholesale Markets Through Market-Based Pricing in Retail Markets." *Electricity Journal* 3, pp. 13–23.

Caves, Douglas, Joseph Herriges, and Kathleen Keuster. (1989) "Load Shifting Under Voluntary Residential Time-of-Use Rates." *Energy Journal* 10, pp. 83–99.

Coyle, Eugene. (2002) "Economist's Stories and Culpability in Deregulation." *Electricity Journal* 8, pp. 90–96.

Doucet, Joseph, and Andrew Kleit. (2002) "Metering in Electricity Markets: When Is More Better?" In *Markets, Pricing, and Deregulation of Utilities*, ed. Michael A. Crew and Joseph C. Schuh, pp. 87–108. Boston: Kluwer Academic Publishers.

Eakin, Kelly, and Ahmad Faruqui. (2000) "Bundling Value-Added and Commodity Services in Retail Electricity Markets." *Electricity Journal* 10, pp. 60–68.

Earle, Robert. (2000) "Demand Elasticity in the California Power Exchange Day-Ahead Market." *Electricity Journal* 8, pp. 59–65.

Edison Electric Institute. (1991) "Historical Background—Electric Utility Industry." at www.eei.org/public/history.htm.

Energy Information Administration, U.S. Department of Energy. (2000) "Historical Overview of the Electric Power Industry." In *The Changing Structure of the Electric Power Industry 2000: An Update*, at www.eia.doe.gov/cneaf/electricity/chg_stru_update/chapter2.html.

Federal Deposit Insurance Corporation. (1997) "The Savings and Loan Crisis and Its Relationship to Banking." History of the Eighties, Volume I: An Examination of the Banking Crises of the 1980s and Early 1990s. Available at http://www.fdic.gov/bank/historical/history/167_188.pdf.

Fox-Penner, Peter. (2001) "Direct Testimony of Peter Fox-Penner on Behalf of Puget Sound Energy, Inc." Washington Utilities and Transportation Commission, November.

Fraser, Hamish. (2001) "The Importance of an Active Demand Side in the Electricity Industry." *Electricity Journal* 9, pp. 52–73.

General Accounting Office. (2004) "Electricity Markets: Consumers Could Benefit From Demand Programs, but Challenges Remain." GAO-04-844, August. Available at http://www.gao.gov/new.items/d04844.pdf.

George, Stephen, and Ahmad Faruqui. (2002) "The Economic Value of Market-based Pricing for Small Consumers." Presentation to the California Energy Commission, March.

Herriges, Joseph, Mostafa Baladi, Douglas Caves, and Bernard Neenan. (1993) "The Response of Industrial Customers to Electric Rates Based upon Dynamic Marginal Costs." *Review of Economics and Statistics* 75, pp. 446–54.

Hirst, Eric. (2001a) "Direct Testimony of Eric Hirst on Behalf of Puget Sound Energy, Inc." Washington Utilities and Transportation Commission, November.

———. (2001b) "Price-Responsive Demand in Wholesale Markets: Why Is So Little Happening?" *Electricity Journal* 4, pp. 25–37.

———. (2002a) "The Financial and Physical Insurance Benefits of Price-Responsive Demand." *Electricity Journal* 15(4):66–73.

———. (2002b) *Reliability Resources of Price-Responsive Demand*. Oak Ridge, TN.

———. (2002c) "Barriers to Price-Responsive Demand in Wholesale Electricity Markets." Edison Electric Institute research paper, June.

———. (2002d) "Demand Response: How to Reach the Other Side." *Electric Perspectives* 27, September/October, pp. 16–33.

———. (2002e) "Reliability Benefits of Price-Responsive Demand." *IEEE Power Engineering Review* 22, November, pp. 16–21.

———. (2003) "Water Heaters to the Rescue: Demand-Side Bidding in Reserve Markets." *Public Utilities Fortnightly* 141, September 1, pp. 32–38.

Holland, Stephen, and Erin Mansur. (2004) "Is Real-Time Pricing Green? The Environmental Impacts of Electricity Demand Variance." CSEM Working Paper 136, August. Available at http://www.ucei.berkeley.edu/PDF/csemwp136.pdf.

Jarrell, Gregg A. (1978) The Demand for State Regulation of the Electric Utility Industry. *The Journal of Law and Economics* 21:269–95.

Lyon, Thomas. (2000) "Capture or Contract? The Early Years of Electric Utility Regulation." Paper presented at 2001 American Economic Association meetings.

McKinsey & Company. (2001) "The Benefits of Demand-Side Management and Market-Based Pricing Programs." May.

———. (2002) "Responsive Demand: The Opportunity in California." Presentation to the California Energy Commission, March.

Michaels, Robert. (2003) "Can RTO Market Merits Be Really Independent?" *Public Utilities Fortnightly* 141, July 1, pp. 35–38.

Mueller, Keith. (2004) "Envision the Utility of Tomorrow." *Public Utilities Fortnightly* 142, June, pp. 62–66.

Neenan, Bernie, Richard Boisvert, and Peter Cappers. (2002) "What Makes a Consumer Price Responsive?" *Electricity Journal* 3, pp. 52–59.

Neenan, Bernard, Donna Pratt, Peter Cappers, Richard Boisvert, and Kenneth Deal. (2002) *NYISO Price-Responsive Load Program Evaluation Final Report.* New York Independent System Operator, January.

O'Sheasy, Michael. (2002a) "Real Time Pricing at Georgia Power Company." Appendix A in Severin Borenstein, Michael Jaske, and Arthur Rosenfeld, "Dynamic Pricing, Advanced Metering, and Demand Response in Electricity Markets." Center for the Study of Energy Markets Working Paper 105, October. Available at http://repositories.cdlib.org/ucei/csem/CSEMWP-105.

———. (2002b) "Is Real-Time Pricing a Panacea? If So, Why Isn't It More Widespread?" *Electricity Journal* 15(10), December, pp. 24–34.

———. (2003) "Demand Response: Not Just Rhetoric, It Can Truly Be the Silver Bullet." *Electricity Journal* 16, December, pp. 48–60.

O'Sheasy, Michael, and Mike Becker. (2002) "Betting on Retail Risk Management: Flat Prices for Peak Hedging." *Public Utilities Fortnightly*, November 1, pp. 28–32.

Rassenti, Stephen, Vernon Smith, and Bart Wilson. (2001) "Turning Off the Lights." *Regulation* 24 (Fall): 70–76.

———. (2002) "Using Experiments to Inform the Privatization/Deregulation Movement in Electricity." *Cato Journal* 21: pp. 515–44.

Swofford, Gary. (2002) Puget Sound Energy Presentation to FERC Demand Response Conference. February.

Taylor, Thomas, and Peter Schwarz. (1990) "The Long-Run Effects of a Time-of-Use Demand Charge." *Rand Journal of Economics* 21, pp. 431–45.

Violette, Dan. (2004) "Evaluation of the Energy-Smart Pricing Plan." Summit Blue Consulting, February. Available at http://www.energycooperative.org/pdf/ESPP-Final-Report.pdf.

Viscusi, W. Kip, John M. Vernon, and Joseph E. Harrington, Jr. (2000) *Economics of Regulation and Antitrust*. Cambridge, Mass.: MIT Press.

Winters, Tobey. (2001) "Retail Electricity Markets Require Marginal Cost Real-Time Pricing." *Electricity Journal* 9, pp. 74–81.

NOTES

1. I am grateful to Michael Giberson, Andrew Kleit, and Alex Tabarrok for helpful comments and suggestions, and to Kristle Kilijanczyk for valuable research assistance.

2. Price might not rise in the less expensive hour because generators are often willing to accept lower prices to avoid having to shut off generators in that hour.

3. The insurance aspects that are built into the costs of maintaining extra plants and equipment are more properly understood as insurance against outage risk, not against price risk.

4. I am grateful to Michael Giberson for pointing out this feature to me.

4

Using Experiments to Inform the Privatization/Deregulation Movement in Electricity

Vernon L. Smith, Stephen J. Rassenti, and Bart J. Wilson

INTRODUCTION: HOW THE COMPUTER-BASED LABORATORY CHANGED THE WAY WE THINK ABOUT EXPERIMENTAL ECONOMICS AND OPENED THE WAY FOR APPLICATIONS

At the University of Arizona, electronic trading (now commonly known as e-commerce) in the experimental laboratory began in 1976 when Arlington Williams (1980) conducted the initial experiments testing the first electronic "double auction" trading system that he had programmed on the Plato operating system. The term "double auction" refers to the oral bid-ask sequential trading system used since the nineteenth century in stock and commodity trading on the organized exchanges. This system of trading has been used in economics experiments since midcentury, and is extremely robust in yielding convergence to competitive equilibrium outcomes (Smith 1962, 1982a). Since information on what buyers are willing to pay, and sellers are willing to accept, is dispersed and strictly private in these experiments, the convergence results have been interpreted (Smith 1982b) as supporting Hayek's thesis "that the most significant fact about this [price] system is the economy of knowledge with which it operates, or how little the individual participants need to know in order to be able to take the right action" (Hayek 1945, 526–27).

As with all first efforts at automation, the software developed by Williams allowed double auction trading experiments that previously had kept manual records of oral bids, asks, and trades to be computerized. That is, it facilitated real-time public display of participant messages and recording of data, and greater experimental control of a process defined by preexisting technology.

It did not modify that technology in fundamental ways. This event unleashed a discovery process commonplace in the history of institutional change: the joining of a new technology to an incumbent institution causes entirely new, heretofore unimaginable institutions to be created spontaneously, as individuals are motivated to initiate procedural changes in the light of the new technology. Electronic exchange made it possible to vastly reduce transactions cost: the time and search costs required to match buyers and sellers to negotiate trades, including agreements to supply transportation and other support services. More subtly, it enabled this matching to occur on vastly more complicated message spaces, and allowed optimization and other processing algorithms to be applied to messages, facilitating efficient trades among agents that had been too costly to be consummated with older technologies. Moreover, resource allocation problems thought to require hierarchical command and control forms of coordination, as in regulated pipeline and electric power networks, became easily susceptible to self-regulation by entirely new decentralized pricing and property-rights regimes. Coordination economies in complex networks could be achieved at low transactions cost by independent agents, with dispersed information, integrated by a computerized market mechanism. This realization then laid the basis for a new class of experiments in which the laboratory is used to test-bed proposed new market mechanisms to enable a better understanding of how such mechanisms might function in the field, and to create a demonstration and training tool for potential participants.

We provide a short history of the application of the conception of smart computer-assisted markets to the design of electricity markets here and abroad.

THE PRIVATIZATION/DEREGULATION MOVEMENT IN ELECTRICITY[1]

The Arizona Utility Study

In 1984 the Arizona Corporation Commission (ACC) contracted with the University of Arizona experimental economics group to study alternatives to rate-of-return regulation of the utilities, with particular emphasis on electric power. The study consisted of two parts: incentive regulation (Cox and Isaac 1986), and deregulation (Rassenti and Smith 1986). (Also see Block et al. 1985.) Only the second part will be discussed here since this was the study that led to a long and continuing research program, encouraged by the privatization movement, with applications first in New Zealand, then Australia, and most recently in the United States.

Recommendations

The deregulation portion of the study led to many detailed recommendations that can be briefly summarized in the following key points (see Rassenti and Smith 1986):

1. The energy (generation) and wires (transmission and distribution) businesses would be separated, with generator companies (gencos) spun off from parent-integrated utilities through the issuance of separate ownership shares to form independent companies.
2. An economic dispatch center (EDC) would be formed that would operate a computerized spot auction market for determining prices and allocations based upon hourly offer price schedules submitted by gencos that were location (node) specific. The spot market should be constituted so as to facilitate and incentivize the eventual inclusion of demand-side bidding by discos (distribution companies and any other commercial and industrial bulk or wholesale buyers). Thus, ultimately and ideally, prices would be determined in an hourly two-sided auction in which discos would submit location-specific bids to buy energy delivered to their location just as gencos would submit offers to inject energy at their respective locations on the grid.
3. Discos and transcos (transmission companies) would not be protected by exclusive franchise permits, and would be subjected to the price discipline of potential, if not actual, entry.
4. Important functions of existing institutions would be preserved but operate through a computerized spot market bidding mechanism based on decentralized ownership of gencos. By "existing institutions" we referred to optimization—historically, computerized dispatch based on the engineering cost characteristics of generators and the network of integrated utilities—joint ownership by utilities of shared transmission capacity, and power-pooling rules for security (spinning) reserves. In the proposed competitive reorganization, optimization algorithms would not be applied to production and transmission "cost" as in the regulated, hierarchical, integrated utility, but to the offer-supply schedules and bid-demand schedules submitted to the computer dispatch center. The algorithms would maximize the gains from exchange (rather than minimize engineering cost as under regulation) in response to the real-time decisions of all buyers and sellers in the wholesale market. This specification was motivated by the recognition that supply cost is subjective and measured by the willingness to accept payment for energy produced on location; and demand is subjective and measured by

the willingness to pay for delivered energy, where both types of information express the particular *real-time* circumstances of individuals. Coordination was a consequence of a new form of property rights: (i) rules for processing messages generated by decentralized agents themselves empowered by rights to choose offers and bids; (ii) contingency rules for accepting offers and bids based on their merit order (higher bids and lower offers have priority in the rank ordering of bids and of offers), but importantly *qualified* by technical and security constraints that are essential if each agent is to bear the *true opportunity cost* that the agent imposes on all others.

The term "property rights" as we shall use it provides a guarantee allowing action within the guidelines defined by the right. Such guarantees are against arbitrary reprisal in that they restrict punitive strategies that can be levied against actions taken by the rights holder. Such guarantees provide only limited certainty of protection. Most specifically, property rights, as a guarantee allowing action, do not guarantee outcomes, since outcomes depend upon the property rights of others, and in electricity markets, as we shall see, global constraints affect local outcomes that must be honored if the system is to be efficient and dynamically stable, and to incentivize the direction and level of capital investment.

Defining Competitively Ruled Property Rights to Unique "Monopolistic" Facilities

It was the ACC project that alerted us to the existence of "cotenancy contracts" for the joint ownership and operation of some large generation and transmission facilities. For us this was an illuminating empirical discovery, since this institution, which we modified with competitive property-right rules, offered the potential to render the concept of natural monopoly *null and void*. Thus, suppose a city demand center can be adequately served by a unique physical facility such as a pipeline or transmission line. Under American-style regulation it is decreed that an exclusive franchise will be awarded to a single owner of the facility, whose price will be set so as to regulate the owner's rate of return on investment. Alternatively, in our proposed competitively ruled joint-ownership property-right regime it is decreed that (1) the facility must have two or more co-owners each having an agreed share of the rights to the capacity of the facility (in practice a common cotenancy contract rule is for each cotenant to receive capacity rights in proportion to his contribution to capital cost). In addition, (2) competitive rules would allow rights to be freely traded, leased, or rented; and (3) the rules would also allow new rights to be created by agreement to invest in capacity expansion by any subset of the co-owners, through unilateral action by any co-owner, or by outsiders if the existing owners resist expansion to meet increased demand. In historical practice, cotenancy contracts had prohibited sale by individual rights holders without the consent of the other cotenants, and capacity expansion was allowed to occur only by joint agreement. The proposed new property-rights structure creates multiple rights holders to compete in marketing downstream services using the unique facility and encourages new investment in response to increased demand. Subsequent to the ACC study, new research uncovered other examples of cotenancy contracts, a common one being the joint ownership of specialized printing facilities by a consortium of newspapers in a city. Clearly, who prints the newspapers is a production issue potentially separable from the competition of newspapers for subscribers and advertising services. The courts repeatedly affirmed this principle when such cotenancy contracts attempted to include marketing and pricing conditions in what was ostensibly a shared-production agreement (Reynolds 1990). Thus our conception of a joint-venture property-right regime had already been well articulated in court cases involving newspapers. There was no new principle, only the question of how it might be reformulated for application to network industries.

This model of cotenancy as an instrument of competition was further elaborated in Smith (1988, 1993), and tested experimentally in the context of a natural gas–pipeline network funded by the Federal Energy Regulatory Commission (Rassenti, Reynolds, and Smith 1994). The model would also play a facilitative role in our consulting on privatization in New Zealand. But such discussions are far from culminating in a completed instrument, with many practical implementation difficulties remaining.[2]

Aftermath of the Arizona Study

By 1985 when the study report was filed and presentations made to the ACC, the political composition of the commission had altered, and the immediate impact of the recommendation for deregulation on Arizona policy was nil. By the time our final report was completed the commission was composed of new elected officeholders, and they considered our proposal to be impractical, idealistic, and politically infeasible. Of course the commission's actions made the last claim a self-fulfilling truth. Subsequent developments would reveal that this experience was a minor battle in a wider war for institutional change that would begin abroad but would ultimately spread to the United States, but with less success, we believe, than abroad.

Contrary to the position of the new commission, we considered our proposal eminently feasible in the electronic age, though in need of *far more fundamental* research, and we resolved to undertake controlled experimental studies of various issues in the deregulation debate. Progress on this objective, however, was slow due to inadequate funding, and the fact that the cost of software development for the laboratory study of electronic trading in the context of electric networks was higher than for traditional forms of experimental research. Nevertheless, by 1987 we had conducted several pilot experiments in a six-node electric power network with three fixed, inelastic nodal demand centers, and nine gencos (described in Rassenti and Smith 1986). The gencos, located at various nodes, submitted sealed-offer price schedules each trading period to supply power over transmission lines whose energy losses were proportional to the square of energy injected. A valuable lesson from this unpublished research was the ease with which gencos could push up prices against inelastic demands by bulk buyers using a mechanism that did not permit demand-side bidding to implement consumer willingness to have deliveries interrupted conditional on price. This was our first brush with the important principle that competition is compromised in supply-side auctions in which buyers are passive and are unable through the mechanism to enter demand-side bid schedules. We report experiments below that provide a rigorous demonstration that when the spot auction mechanism in common use around the world is supplemented by demand-side bidding, it provides a property-right regime that is an effective antitrust remedy.

Domestically, through the 1980s and into the 1990s, electric power would remain subject to American-style rate-of-return regulation, while abroad government-owned electric (and other) utilities were under political pressure to explore the use of markets for electrical energy allocations. Industry performance was seen as abysmal in the 1980s, causing countries such as Chile, the United Kingdom, and New Zealand to think the unthinkable: decentralization might be preferable to either government planning or direct regulation. But how might it be done?

HOW EXPERIMENTS WERE USED TO INFORM PRIVATIZATION: NEW ZEALAND AND AUSTRALIA

Beginning in 1986 we initiated software development and a series of experiments to study mechanism design, industry structure, pricing, transmission, and market-power issues in electricity markets (Rassenti and Smith 1986; Backerman, Rassenti, and Smith 1997; Backerman, Denton, Rassenti, and Smith 1997; Denton, Rassenti, and Smith 1998; Rassenti, Smith, and Wilson 2000). While this research was proceeding, two of the authors (Rassenti and Smith) consulted for the New Zealand Treasury in 1991 and 1993, and for Australia's Prospect Electricity in 1993 and National Grid Management

Council in 1994. The impetus in New Zealand was our 1985 ACC report that fell unceremoniously on deaf ears in Arizona but attracted attention abroad.

What Were the Questions?

The following research questions, addressed in laboratory electricity-network experiments after 1986, and motivated by our ACC study, provided the primary information base for informing our contribution to the privatization process in electricity down under.

> (1) Is decentralization feasible and, if so, is it efficient to combine decentralized property rights in energy supply with a computer-coordinated spot market and optimization schemes for dispatching generators?

Before the first experimental observations were made it was an open question whether it was feasible to replace engineering cost minimization in large, integrated utility hierarchies with independent gencos submitting node-specific asking-price schedules, bulk buyers submitting node-specific bid-price schedules, and allocations determined by algorithms maximizing the gains from exchange implied by these marginal bid/ask schedules and the physical characteristics (loss characteristics and capacity constraints) of the grid. Engineers and managers to whom we made presentations were overwhelmingly skeptical—in fact were openly hostile—that such a system could be relied upon. The conventional wisdom of economists had been stated as follows:

> Generation and transmission are intimately and fundamentally related by the interconnections that the transmission system provides and the associated opportunities for area wide optimization. . . . Because of these relationships, decisions either short-run or long-run, made at any point in a power system affect costs everywhere in the system. These effects raise potential externality problems. If a power system's components are owned by more than one firm, it is crucial for the efficiency of short-run and long-run decision making that all owners of parts of the system take into account all effects of their actions, not just the effects on the part of the system they own. (Joskow and Schmalensee 1983, 63)

Experimental markets, in which all energy sales and purchases were expressed as offers to sell and bids to buy so that allocations were determined simultaneously given the physical properties of the grid, demonstrated that energy market deregulation was eminently feasible. Furthermore, short-run efficiency was high—on the order of 90–100 percent of the maximum economic surplus, or gains from exchange, were achieved. Figure 4.1 shows a plot of efficiency for two experimental sessions consisting of a series of thirty

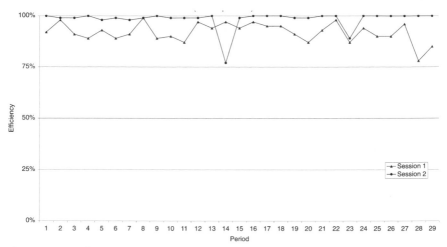

Figure 4.1.   Efficiency with Other-Side Rule and Unconstrained Transmission Line (Twice Experienced)

trading periods using experienced subjects in a three-node radial network (Backerman, Rassenti, and Smith 1997). Why are there no important efficiency losses due to short-run externalities? The answer resides in the condition that all allocations are determined simultaneously. Power loss on shared transmission lines varies as the square of total power injected. Therefore, genco A suffers higher costs of energy loss if genco B is using the same line. But if optimization is based upon every agent's marginal willingness to pay or to supply, with price and allocations determined simultaneously, then each agent bears the appropriate *opportunity* cost that his action imposes on all others at the margin. The problem is solved by the simultaneous submission of bid/ask schedules to which are applied algorithms for maximizing the implied gains from exchange, taking account of system transmission losses.

But there are many other potential "external effects," besides shared system energy losses, that in principle are or can be internalized via mechanisms that link bid/ask schedules with system constraints through rule-governed coordination: (i) voltage "constraints" (as they are so treated, technically, in all operating systems today), requiring "reactive power" to be produced, and therefore priced in the market if such constraints are to be incorporated into the market process;[3] (ii) intertemporal links on both the demand and generator sides of the market historically have implied the need for optimization over time, not just in the current spot market, but as shown by Kaye and Outhred (1989) and Kaye, Outhred, and Bannister (1990) the primary intertemporal coordination requirements can be met by futures markets; (iii) contingency provisions such as generator and transmission reserves to avoid blackouts from unscheduled equipment outages, and to avoid unstable cascades of outages that spread through the network.[4]

(2) How is the answer to question 1 affected by demand-side bidding?

Both regulation and government ownership have produced industries with a strong supply-side orientation. The politics of power yields a system in which (i) there are severe political repercussions if consumers "lose lights" too often, and (ii) consumers making decisions have no means of directly (or indirectly through wholesale markets) comparing the cost of new capacity with the cost of interruptions on-peak or in emergencies. Consequently, adequate reserve capacity in generation and transmission requires supply-side investment sufficient to meet all demand, plus a large margin for security of supply. The regulatory and government-owned systems had no incentive to install technologies for relieving load stress by introducing time-of-demand pricing and voluntary interruptible contracts for customers. For this to occur, power users must have the real-time spot market capacity to either directly reduce consumption in response to price increases or indirectly by contract with the distributor to effect reduced deliveries in response to price increases. As we shall see below, the capability for interruption of energy flows must be expressible in the spot market if prices are to be adequately disciplined.

New Zealand

Our consulting work in New Zealand was directed entirely to questions of how a privatized NZ electrical industry, and a wholesale power market, might be structured. Intellectually, the sea change in issues of privatization versus government ownership and regulation was so drastic in the direction of economic liberalization in the early 1980s, after the election of a new reform-committed Labour government, followed by a foreign-exchange crisis the next day, that electricity reform seemed certain. All government enterprises had performed so poorly and were such a drain on the treasury that the country was soured on the "NZ (socialist) experiment." Everywhere in New Zealand, by 1991, were to be found people expressing the "user pays" principle as a slogan of reform.[5] This exuberance, strong in the late 1980s and early 1990s, is now much abated. "New Zealand . . . retains large state-owned corporations that are suitable for privatization, but . . . its privatization activity has been muted for much of the 1990s. This decline reflects political perceptions of the privatization act as well as the resolution of property right issues, some of which arise from considerations of industry structure that is suitable for light-handed regulation, and some from the potential settlement of Maori claims on the crown" (Evans 1998, 3).

Our consulting for the New Zealand Treasury (1991), and later, Transpower, NZ (1993), created as the State Owned Enterprise that maintained and operated the high-voltage grid, emphasized privatizing transmission, transmission pricing, and demand-side bidding.

Privatizing Transmission

What might be the incentive and ownership structure that should be implemented for the New Zealand grid and for the market dispatch center that would determine allocations of energy supply among decentralized generator owners who bid into the spot market?

Our recommendations had their genesis in our 1985 ACC study of cotenancy, but the basic idea—a cotenancy property-right system—was substantially extended and tailored to fit the special physical properties of electric power flows in interconnected alternating current (AC) networks. Primarily these properties are twofold: (i) flows on individual links in the network cannot be precisely controlled because in AC networks there has not existed anything analogous to the valves on links in fluid and gas pipeline networks; (ii) optimization in such networks requires knowledge of willingness-to-pay bid-demand values at delivery nodes, offer-supply terms at power injection nodes, and the physical properties (loss characteristics and capacity constraints) of all elements of the network. One can then solve simultaneously for the pattern of energy injections and deliveries that satisfies all demands and constraints while maximizing the short-run gains from exchange based on all such information. These two characteristics combined imply that it is not possible to specify well-defined path rights from any source node to any delivery node. The flow on a given path may be optimal at one time, but with a change in the pattern of supply and demand, and with different transmission constraints binding, the flow on that path may be much different, even reversed.

We proposed that these characteristics of the electricity industry be supported by a property-right regime with the following commensurate features when the system is privatized as a joint (competitively ruled) venture, or cotenancy, owned by all users.

1. At each energy injection node is connected a set of generators with some specified capacity that has occurred in history up to the time of privatization. That capacity is assumed to reflect the benefits, based on historical utilization rates, and site value of locating the capacity where it resides.
2. Similarly, each delivery node will have associated with it a capacity to withdraw power.

3. Rights to inject (or withdraw) power at each node can then be defined and certificated in capacity terms based on historical investment.
4. Each generator has the right to submit a bid-supply schedule indicating the various quantities the supplier is willing to inject at corresponding stated asking prices, where the schedule is restricted not to exceed a total offer of that generator's capacity rights at its connection node. How much of this offer is accepted by the dispatch center depends on the offer terms of competing suppliers at the same or other nodes, the nodal pattern of demand, and the physical properties of the grid at any time. Thus, each generator merely has a right to offer up to its capacity in units of power, not the right for the offer to be accepted.
5. These capacity rights can be freely traded, leased, or rented to others subject only to contract laws applicable to any industry.
6. Any individual user in this structure, or any group of users forming a consortium, is free to invest in increasing the capacity of any line or lines in the system. Those making the investment will acquire rights, as in 3–5 above, to any increase in capacity at individual nodes that is made possible by the investment. Any such increases in capacity will be uncertain and based on engineering simulations that are commonly used to evaluate capacity expansions.
7. Finally, since incumbent users may not be well motivated to expand capacity, the cotenants cannot prevent the entry of new investors who invest in line capacity expansion and acquire rights to the consequent increase in nodal rights to inject (or withdraw) power.

Transmission Pricing

Given the joint-ownership structure indicated above, all users share output-invariant operating and maintenance costs in proportion to their respective capacity rights. The primary variable cost of transmission is the energy lost in the transfer of power from source nodes to delivery nodes. This loss (per mile of line) varies approximately as the square of energy injected—less energy is received than is sent. Hence, if the per-unit average loss is A, for a given line, the marginal loss is $M = 2A$. This implies that if the price at an upstream injection node is P, then at any downstream node the price is $P' = P + M$; that is, the delivery price is the price at the injection point plus the marginal cost of energy lost in delivery. Note that M is the true opportunity cost of energy lost in transmission, and all buyers served by remote generators must pay this cost in an efficient energy supply network. On long lines, where the average loss at peak demand can be 15–20 percent, the price difference, $P' - P = 2A$ can be 30–40 percent of the delivered price.[6]

Demand-side Bidding

Competition is greatly enhanced if wholesale buyers can bid into the spot market using discretionary demand steps that define price levels above which they are prepared to interrupt corresponding blocks of power consumed. As we shall see below, demand-side bidding also reduces price spikes on-peak. Moreover, interruptible flows substitute for security reserves of generation capacity, while reducing the prospect that transmission lines will become constrained.

New Zealand deliberations on structuring the grid continue. However, the functions of the spot market, called the New Zealand Electricity Market (NZEM), have been structured as a rule-governed joint venture. (For a detailed report see, Arnold and Evans 2000; also see NZEM 1999.) Only three countries have implemented policies requiring the grid users to fund investment expansions: Chile, Peru, and Argentina. In all three cases, however, the multiple owners operate under regulated prices (Kleindorfer 1998, 69). Thus no country has implemented a completely privatized grid regulated only by property rights.

Although our fledgling proposals for structuring joint ownership of the grid have not been implemented, and indeed require a lot more intensive work to be operational, the New Zealand spot market implements both the marginal-loss pricing of transmission and demand-side bidding. It is important, however, to note that nodal energy pricing in New Zealand does not provide *ex ante* real-time prices that can be avoided by action of buyers and sellers in the current period. Prices are an *ex post* cost-recovery and distribution scheme, and affect decisions only insofar as events/conditions are repeated and anticipated by decisions. The same is true for the systems implemented in California and the Middle Atlantic regions in the United States. This is partly the result of industry traditions that think of prices as cost-recovery devices rather than signals of avoidable opportunity costs and partly a consequence of implementing the appropriate technology. New Zealand, however, is moving to implement true avoidable cost pricing as used now in Australia (see below).

Marginal-cost pricing of transmission is politically very difficult to implement in democratic regimes—three other countries (Chile, Peru, and Australia) have adopted it (Kleindorfer 1998, 69). Strong political pressures favor averaging transmission losses across all customers, which, let it be noted, creates an incorrectly priced external effect that is avoidable by appropriate specification of property-right rules, and illustrates one of the many externality problems created by collective action. With minor exceptions, averaging losses over all customers was the universal practice in both state-owned and American-style regulatory regimes, and this practice dies very hard. People do not understand the opportunity cost-efficiency principle here: each agent pays the cost that his consumption imposes on others, thereby eliminating external effects. But collective agreement is necessary to implement the application of this principle to grid pricing. (Note that the principle creates no problem in the airline or accommodation industries, where on-peak prices emerge spontaneously in competition, à la Hayek's (1945) perceptive argument, and collective agreement is not needed. This illustrates one of the many hazards in privatizing interdependent network industries using some collective agreement process.)

Most of the New Zealand population and electricity demand is on the North Island, while most of the generation capacity is on the South Island. It is nine hundred miles from the bottom of the South Island, where the most remote generators are sited, to the top of the North Island, where the largest concentration of population is located (Auckland). Consequently, at peak demand, with no constrained lines causing a further price difference due to congestion, there is a price difference of approximately 35 percent between the two most remotely separated nodes. Figure 4.2 provides a chart of NZ electricity prices at the terminal stations of the interisland link (not the two extreme nodes), Haywards in the North and Benmore in the South, for the winter months of July and August, when the heating demand for energy is greatest.

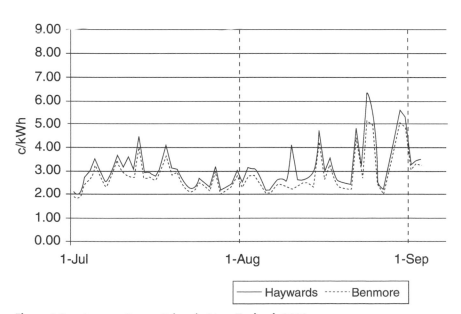

Figure 4.2. Average Energy Prices in New Zealand, 2000

Relevant to demand-side bidding, the New Zealand Electric Market (NZEM) rules specify that "Each trading day, each Purchaser Class Market Participant will submit to the Scheduler the bids pursuant to which . . . [that Participant] . . . is prepared to purchase Electricity from the Clearing Manager for each trading period of the following trading day" (NZEM B.2.1). Such bids specify the relevant trading periods, the grid exit node, must represent reasonable endeavors to predict demand and specify up to ten prices (price steps or "bands") and corresponding quantities. There are no upper or lower limits on prices. "The highest price band for each bid will be deemed to start at a quantity of zero" (NZEM 1999, B.2.3). Note that this provision defines the strike price where the Marshallian demand schedule intersects the price axis. Since the technology for interrupting flows is limited, these provisions of the NZEM are currently little used (as reported to us in private conversation with Lewis Evans at Victoria University, NZ), but the institutional stage is set for more extensive demand-side bidding as the appropriate technology becomes more available and cheaper. They will become more significant when New Zealand implements real-time pricing.

Australia

We were invited to visit Australia in 1993 by Prospect Electricity (now part of Integral Energy) in New South Wales, the second largest distribution company in that state. Australia, unlike New Zealand, was not committed to privatizing electricity, although the political debate had begun. Rather, the commitment was to decentralization; that is, setting up a national wholesale market. This was the charge of the National Grid Management Council (NGMC). (Privatization if it occurred was the province of the states, which were the owners of existing power system assets. All generation, transmission, and distribution systems remain publicly owned even today, with the exception of Victoria, where all are privately owned, while South Australia has executed two-hundred-year leases of its assets to private entities.)[7] It was during this visit that we learned that the constituency for privatization was made up of bulk buyers—commercial, industrial, and distribution companies—who believed that the state government–owned electricity industries were producing power at exorbitant cost, and this was hampering the ability of Australia's energy-intensive industries to compete in world markets. Primarily our sponsors consisted of the buyer side of the industry, and our task was to supply market information and demonstration technology: give lectures and seminars and conduct experimental workshops with a wide spectrum of industry and government representatives in which they would participate in our prototype wholesale electricity experiments, demonstrating feasibility, efficiency, and possible structural features for a decentralized wholesale market, with the extent of privatization yet to be determined. These lectures and workshops were well-attended, but with understandably more enthusiasm coming from the demand side than the supply side. Such was the political environment.

Subsequently, the central government created the National Grid Management Council to oversee, and plan, a wholesale energy market embracing the states, integrated by a national interconnected grid. This led to a controversial "paper trial" (cost, $2 million) in which participants walked through proposed procedures for bidding and clearing in a spot energy market. Our Australian contacts pressed, and won, approval to conduct laboratory experiments with a prototype for the proposed market. We were consultants on software specifications and experimental design, with all development and experiments to be conducted in Australia. This ultimately led to a two-week (seven hours per day) electronic trading experiment using nonindustry participants trained in the exchange procedures, and earning profits based on induced costs, and demands, using Australian parameters and grid characteristics. We advised against using any industry participants because of their known political biases for or against the market reforms.

On December 13, 1998, the National Electricity Market began trading Australian electricity. Prior to that period separate markets traded power in the states of Victoria and New South Wales as early as 1996.

In summary, experimental methods in economics served to facilitate the development of a wholesale electricity market in Australia in the following ways:

1. It provided a database demonstrating the feasibility of using a smart market and price signals to coordinate production and transmission over huge geographical areas, and to help inform the political decision process.
2. Treatment results from specific experimental designs suggested that overall market efficiency, price volatility, and the distribution of surplus among the buyers, sellers and the transmission system were significantly impacted by the following: transmission and auction market pricing rules, whether or not there was demand-side bidding, and whether or not transmission line constraints were binding.
3. As noted in communication with Hugh Outhred, the experiments "at UNSW also demonstrated the importance of forward markets in containing market power" (see Outhred and Kaye 1996).
4. It provided hands-on experience and training for managers and technical staff, and alerted the principal agents involved in the wholesale market to some of the potential design issues in the process.
5. It enabled the Australians to go through the process of market prototype software development, to conduct experiments using Australian grid

and generator cost parameters, and to learn much more about how their proposed market system might work prior to actual trading in Victoria and New South Wales.

The wholesale market in Australia has implemented features that make it among the most advanced anywhere from the perspective of reflecting good economic design principles, although it is important to emphasize that those principles are under ongoing review and modification in the light of changing experience and technology. We mention two features central to the issues discussed above that were in the National Electric Code prior to their experiments (quoted from personal correspondence with Hugh Outhred, February 2, 2001):

> Network pricing in Australia does incorporate marginal network losses in the following manner: the "notional interconnectors" between regions include . . . [adjustment for] . . . marginal losses . . . directly into the process for setting five-minute prices; interregional transmission loss factors are set annually on the basis of average marginal network losses [the averaging period may be shortened at some future time]. . . .
>
> The Australian National Electric Market Rules (NEM) . . . [also] incorporate the demand side—both formally as bids . . . and informally as price elasticity. The latter option exists because: half-hourly prices are forecast at least 24 hours ahead and broadcast to all market participants (supply and demand side); participants can change their bids and offers from the time of their original submission (one day ahead) down to the half hour to which they apply; the actual spot price is set in "real time" and broadcast to all participants—a consumer can simply reduce demand in response to that price signal and thus avoid paying the price. That facility is now being used in practice, both by a consumer participating directly in the NEM and by retailers backed up by discretionary demand reduction contracts with final consumers." It is evident, however, that "much more development (is) needed." (Outhred 2001, 20)

THE UNITED STATES

Background

The deregulation of electricity did not impact the United States until privatization was well advanced abroad. Viewed from the perspective of those of us interested in market design for deregulation, the U.S. experience has been disappointing, and the design details heavily politicized. For starters, the industry strongly opposed deregulation. Nothing new here; the same was true for airline, railroad, and trucking deregulation. But with electricity there was the need for *collective agreement* on how the industry would be restructured, and what rules would govern market operation since there was clear need for computer coordination of generator loads to meet instantaneous demand on highly interconnected networks. (No need for such agreement in the deregulated airline industry. The routes no longer had to be certificated, the industry was regulated by free entry and exit, and what emerged spontaneously in response to the demand for frequent low-cost service was the hub-and-spoke structure that was anticipated and deliberately planned by no one.) Originally, for example, circa 1985 when we finished our ACC report, the industry had argued that deregulation was technically infeasible, but that proposition had been shot down all over the world by privatization programs, none of which had followed American-style rate-of-return regulation. There were various forms of "light-handed" regulation, such as price caps on charges for the "wires" business—high-voltage transmission or local low-voltage distribution—but energy was more and more priced competitively, limited only by technology and learning. No one abroad wanted to use the American model, which was perceived as broken just as badly as the state-owned or -dominated models that were being discarded.

In this environment, once the writing was on the wall, the utilities focused not on questions of market design and efficient spot markets, but on lobbying for fixed new monthly charges to cover their alleged "stranded costs." This was price design for revenue protection, not market design. Most economists seemed to accept the need for such compensation, either because it was "fair" for utilities to recover the cost of investments made in good faith under a regulatory regime that was being replaced (Baumol and Sidak 1995), or because it was considered the political price to be paid for utility support for deregulation (Block and Lenard 1998). Since the utilities were already privately owned, had long engaged in bilateral economy energy exchanges, and energy marketers or intermediaries had emerged to facilitate such contracts, there was opposition to the very idea of an open spot market. Ironically, the bilateral special interest groups had been fostered by legislation intended to move the industry toward market liberalization: the Public Utility Regulatory Policies Act of 1978, and the Energy Policy Act of 1992. These initiatives were designed to facilitate transmission access by independent power producers as a step toward fostering the development of wholesale power markets. (Bear in mind that such access was being opposed by the utility.) The bilateral trading model was promoted, partly because of its perceived success in reforming the gas industry. California followed this model in restructuring electricity. We long regarded this model as misguided: bilateral bargaining in the electronic age could not provide the foundation for an efficient market model of interdependent (pipeline or transmission) networks.[8]

Thus, using California as an example, the new "wires" utilities succeeded in instituting new fixed monthly charges to cover their stranded costs, and fixed per-unit energy charges for retail customers, but no one was considering demand-side bidding as an instrument to discipline prices in the hourly spot market and to provide incentives for users to reduce demand or switch their time-of-day consumption from higher- to lower-cost periods. Imagine what would be the consequences to the airlines, and all of its passengers, if fliers were charged monthly access fees and a fixed price per mile traveled *independent of flight time of day, time of week, season, or holidays!*

Figure 4.3 illustrates a typical twenty-four-hour period of price variation on the California Power Exchange (PX) (its open spot market exchange). Since most of the power is traded by contracts at secret prices, with only physical characteristics such as the origin, destination, and physical quantities reported for dispatch, the PX prices are representative only of what passes through the hourly auction. Observe in figure 4.3 that the peak demand and most of the "shoulder" transition demand (between peak and off-peak) are at prices above 10 cents per kilowatt ($100 per megawatt), and are therefore far in excess of what local distributors collect from their residential customers. There are numerous other examples of on-peak price spikes of up to one hundred times the normal energy prices (in the $25–$30 per megawatt range). (See the *Bloomberg Daily Power Report* [1999] for a report on sharp price spikes in the Midwest and South.)

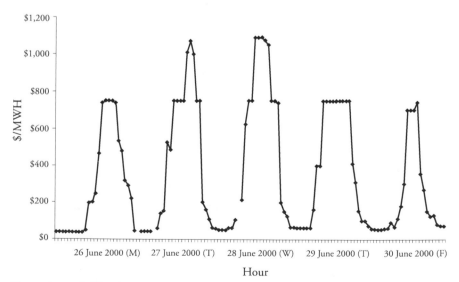

Figure 4.3.   California PX Prices

DEMAND-SIDE BIDDING CONTROLS, MARKET POWER, AND PRICE SPIKES

Earlier experimental market research cited above used demand-side bidding, and we observed very competitive results. New experiments study this issue much more systematically in the design reported by Rassenti, Smith, and Wilson (2000) comparing prices with and without demand-side bidding. Bulk buyers submit discretionary bid steps reflecting the prices above which they are prepared to reduce demand by invoking their contracts for interrupting deliveries. It is important in a competitive electricity market that bulk energy providers contract for discretionary interruption of (suitably compensated) consumers. Why? Because then their bids in the wholesale market cannot be known with certainty by the supply-side bidders, and demand-side bidding can better deter supply-side market power. The problem created by unresponsive demand in a supply-side auction can be illustrated with the chart shown in figure 4.4, provided by Hugh Outhred (University of New South Wales, School of Electrical Engineering). In such a market, the clearing price is sensitive to the asking prices submitted by peaking generators in short supply, especially near peaks in demand. Thus, in figure 4.4, the price is $15 per MW, with demand 7700 MW, but if demand had been 8000 MW, the spot price would have been $45 per MW, and at a demand level of 9000 MW, the

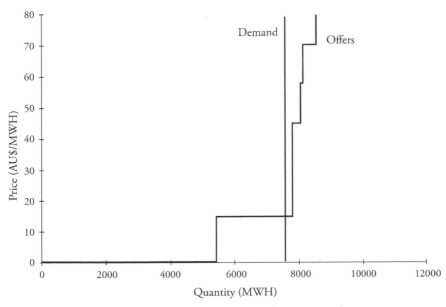

Figure 4.4. Price Determination in the Australian Electricity Market

price would have been indeterminate, forcing the dispatch center to use security reserves or to involuntarily interrupt customers. Unquestionably, some consumers would have been prepared to reduce demand to avoid such a price spike, provided that they had been given the opportunity and incentives commensurate with the savings. Is this to be judged a problem in supply-side market power or an institutional failure of the market mechanism to implement responsive demand? The tendency is to blame market power, although in another industry—hotel/motel accommodations, or airline seat pricing, where the product also is nonstorable—demand is responsive to time variable competitive prices.

Figure 4.5 plots experimental data comparing prices with and without demand-side bidding over the course of fourteen "days" of trading. Each day in an experiment consists of a cycle of four demand-pricing periods: shoulder, peak, shoulder, and off-peak. Hence, the experiments consolidate the shoulder transitions, peak, and off-peak hours (shown in figure 4.3) into four simpler time blocks for auction price determination. Note that when there is no demand-side bidding, prices are much increased, well in excess of the controlled experimental competitive prices, especially on the shoulder and peak demand periods. Both generator "market power" and upward price spikes are effectively controlled by the introduction of demand-side bidding, leaving all other features of the market unchanged. In these experiments a very modest proportion (16 percent) of peak demand is interruptible by wholesale buyer/distributors; most of the on-peak demand (84 percent) is what the industry calls firm or "must-serve" demand.

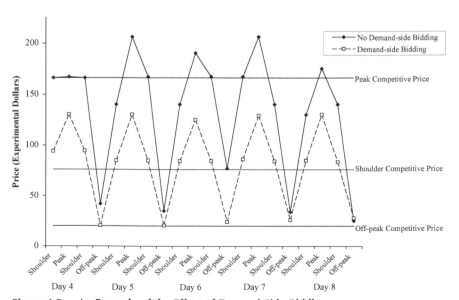

Figure 4.5.   An Example of the Effect of Demand-Side Bidding

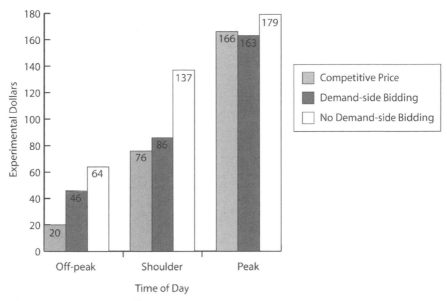

Figure 4.6a. Average Prices

The chart in figure 4.5 plots the data from just one of four independent experimental comparisons reported in Rassenti, Smith, and Wilson (2000). Figure 4.6 provides a bar graph summarizing all of the experimental results. With demand-side bidding the average level of prices is much reduced in all segments of the daily demand cycle, while the great variability in price changes is nearly eliminated.

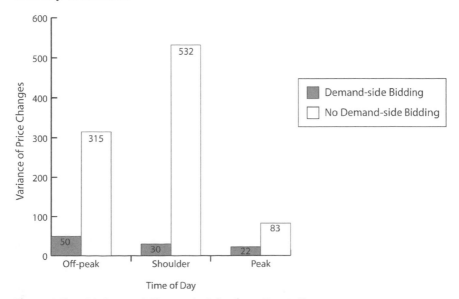

Figure 4.6b. Variance of Changes in Price from Day to Day

SUMMARY AND IMPLICATIONS FOR ELECTRICITY DEREGULATION IN THE UNITED STATES

The computerization of laboratory market experiments using profit-motivated human subjects in the 1970s revolutionized our thinking about the purpose and uses of experiments. In particular we soon came to recognize that the laboratory could be used to test-bed new electronic trading systems for application to industries traditionally perceived as requiring hierarchical organization and government regulation to achieve proper coordination and control over the legally franchised monopoly. Electricity was a prime example, and we attempted to use our first experience with what we called "smart computer-assisted markets" to inform Arizona's cautious and tentative interest in restructuring its electrical industry to rely on markets to regulate the energy segment of the industry. Failing at the time to influence policy, our effort was not ignored abroad, and we participated as consultants in developing proposals and the use of experiments to help inform some of the key research issues in privatization and to serve as a hands-on training tool for those managing the transition. Privatization required the creation of new property rights: an ownership structure for the grid, generator entry and exit rules, market rules governing messages and contracts in the context of computer-controlled coordination, optimization, and communication, but with all outcomes driven by the decisions of dispersed agents whose circumstances of time and place were reflected in market bids to buy or offers to sell.

In the United States the industry was already privatized, but subject to state and national price regulation based on a "fair" return on investment. With the deregulation of electric utility prices and consumption, each state or region needed to develop a plan for restructuring its industry and specifying the auction market rules for determining the real-time wholesale price of energy. Without exception, the resulting market designs, hammered out by regulators, consultants, industry representatives, and various power-marketing intermediaries, all employed supply-side bidding mechanisms for the hourly spot market. These spot markets were supplemented with wide-ranging freedom for power users, producers, and intermediaries to engage in a variety of bilateral contracts outside of direct discipline by the spot market. For the spot market this meant that any user, regardless of the individual circumstances of that consumer's need for an uninterruptible flow of energy, would be guaranteed that his demand would be served. Bilateral contractors could agree to allow various degrees of firmness of demand to impinge on contract terms. But in this longer-term contract, market prices are negotiated and secret and are not subject to the direct real-time opportunity cost constraints provided by the spot market.

The "must-serve" demand policy in the spot market was inherited from a rigid regulatory regime that politicized the reliability of electricity flows to all consumers, whatever the cost. This cost was collectivized by averaging it across all users regardless of individual consumer differences in willingness-to-pay for keeping the lights on. The local utility was expected to maintain service, or restore it quickly, even in inclement weather, spreading the cost of this super-reliability thinly over all customers. This cost included the maintenance of substantial reserves in generation and transmission capacity. Thus system reliability and the capacity to satisfy all retail demand were exclusively a supply-side adjustment problem. The consequence of this supply-side mind-set was uncontrolled cost creep that increased to a gallop and ultimately became part of the political outcry for deregulation. Implicitly, however, the process of deregulation assumed that this built-in supply-side bias did not require fundamental rethinking when it came time to design spot markets for the new world of competition. As always in market institutions, the devil was in the details.

Beginning three years ago in Midwestern and Eastern markets, peak prices hit short-run levels of one hundred or more times the normal price level of $20–$30 per megawatt-hour. This was the predictable direct consequence of *completely unresponsive spot demand impinging on responsive discretionary (bid) supply*. More recently the California spot market has been plagued by exorbitant increases in prices as illustrated in figure 4.3. This has led to political action to impose price caps on this market, which, of course, can only discourage a positive supply response to the shortages. The move to replace American-style regulation with what may become known as American-style deregulation is in danger of being derailed by these interventions.

Controlled comparisons between markets with and without demand-side bidding, in which only 16 percent of peak demand can be voluntarily interrupted, show that the effect of demand-side bidding can dramatically lower both the level of prices and their volatility.

The public policy implications are evident: wholesale spot markets need to be institutionally restructured to make explicit provision for demand-side bidding. Distributors need to incentivize more of their customers to accept contracts for voluntary power interruptions. Industrial and commercial buyers who already have the capacity to handle interruptible energy supply, but who contract outside the spot market, need adequate incentives to participate in the spot market where their more responsive demands can impact public prices. Distributors stand to gain by interrupting demand sufficiently to avoid paying higher peak and shoulder spot prices, and these savings can be used to pass on incentive discounts to customers whose demand, or portions of it, can be reduced or delayed to off-peak periods when supply capacity is ample. In California, news reports indicate that distributors have lost $6 billion buying high (figure 4.3) and selling at vastly lower residential rates.

The technology and capacity for implementing such a policy already exists and can be expanded. This policy recognizes that adjustment to the daily, weekly, and seasonal variation in demand, and to the need to provide adequate security reserves, is as much a demand-side problem as it is a supply-side problem. The history of regulation has created an institutional environment that sees such adjustment as exclusively a supply responsibility. The result is an inefficient, costly, and inflexible system that has produced the recent price shocks and involuntary disruption of energy flows. Demand-side bidding coupled with the supporting interruptible-service incentive contracts can eliminate price spikes and price increases and reduce the need for reserve supplies of generator capacity and transmission capacity.

REFERENCES

Arnold, T., and L. Evans. 2000. "The Law and Economics Basis of Enforcement of Governance of Private Joint Venture Networks: The Case NZEM." New Zealand Institute for the Study of Competition and Regulation.

Backerman, S., S. Rassenti, and V. Smith. 1997. "Efficiency and Income Shares in High Demand Energy Network: Who Receives the Congestion Rents When a Line Is Constrained?" *Pacific Economic Review*, vol. 5, issue 3, pp. 331–47.

Backerman, S., M. Denton, S. Rassenti, and V. Smith. 2001. "Market Power in a Deregulated Electrical Industry." *Journal of Decision Support Systems*, vol. 30, issue 3, pp. 357–81.

Baumol, W., and J. G. Sidak. 1995. *Transmission Pricing and Stranded Costs in the Electric Power Industry*. Washington, DC: The AEI Press.

Block, M., J. Cox, R. M. Isaac, D. Pingry, S. Rassenti, and V. Smith. 1985. "Alternatives to Rate of Return Regulation." Final Report of the Economic Science Laboratory, University of Arizona, to the Arizona Corporation Commission, February 15.

Block, M., and T. Lenard. 1998. *Deregulating Electricity: The Federal Role*. Washington, DC: Progress and Freedom Foundation.

Bloomberg Daily Power Report. 1999, online, summer.

Cox, J., and R. M. Isaac. 1986. "Incentive Regulation." In *Laboratory Market Research*, ed. S. Moriarity, pp. 121–45. Norman: University of Oklahoma Press.

Denton, M., S. Rassenti, and V. Smith. 2001. "Spot Market Mechanism Design and Competitivity Issues in Electric Power." Proceedings of the 31st International Conference on System Sciences: Restructuring the Electric Power Industry. *Journal of Economic Behavior and Organization*. vol. 44, issue 4, pp. 435–53.

Evans, L. 1998. "The Theory and Practice of Privatisation." Victoria University of Wellington working paper, August.

Forsythe, R., and R. M. Isaac. 1982. "Demand-Revealing Mechanisms for Private Good Auctions." In *Research in Experimental Economics*, vol. 2, ed. V. Smith, pp. 45–61.

Grether, D., R. M. Isaac, and C. Plott. 1989. *The Allocation of Scarce Resources*. Boulder, CO: Westview Press.

Hayek, F. A. 1945. "The Use of Knowledge in Society." *American Economic Review* 35, pp. 519–30.

Ishikida, T., J. Ledyard, M. Olson, and D. Porter. 2001. "The Design of a Pollution Trading System for Southern California's RECLAIM Emission Trading Program. *Research in Experimental Economics*, vol. 8.

Joskow, P., and R. Schmalensee. 1983. *Markets for Power*. Cambridge, MA: MIT Press.

Kaye, R. J., and H. Outhred. 1989. "A Theory of Electricity Tariff Design for Optimal Operation and Investment." *IEEE Transactions on Power Systems* 4 (2): pp. 46–52.

Kaye, R. J., H. Outhred, and C. H. Bannister. 1990. "Forward Contracts for the Operation of an Electricity Industry under Spot Pricing." *IEEE Transactions on Power Systems* 5 (1): pp. 606–13.

Kleindorfer, P. 1998. "Ownership Structure, Contracting and Regulation of Transmission Services Providers." In *Designing Competitive Electricity Markets*, ed. H. Chao and H. Huntington. Boston: Kluwer Academic Publishers.

McCabe, K., S. Rassenti, and V. Smith. 1989a. "Designing 'Smart' Computer Assisted Markets in an Experimental Auction for Gas Networks." *European Journal of Political Economy* 5, pp. 259–83.

———. 1989b. "Markets, Competition, and Efficiency in Natural Gas Pipeline Networks." *Natural Gas Journal* 6, pp. 23–26.

———. 1990. "Auction Design for Composite Goods: The Natural Gas Industry." *Journal of Economic Behavior and Organization*, September, pp. 127–49.

McMillan, J. 1998. "Managing Economic Change: Lessons from New Zealand." Graduate School of International Relations and Pacific Studies, University of California, San Diego, July 5.

New Zealand Electricity Market. 1999. *Rules of NZEM*. Wellington, NZ: The Marketplace Company Ltd.

Olson, M., S. Rassenti, and V. Smith. 2001. "Market Design and Motivated Human Trading Behavior in Electricity Markets." *IIE Transactions on Operations Engineering*.

Outhred, H. 2001. "Electricity Industry Restructuring in California and Its Implications for Australia." Report for the National Electricity Market Management Company, February.

Outhred, H., and R. J. Kaye. 1996. "Structural Testing of the National Electricity Market Design: Final Report." Report for the National Grid Management Council, Unisearch, September.

Rassenti, S. "0-1 Decision Problems with Multiple Resource Constraints: Algorithms and Applications." Unpublished Ph.D. thesis, University of Arizona, 1981.

Rassenti, S., S. Reynolds, and V. Smith. 1994. "Cotenancy and Competition in an Experimental Auction Market for Natural Gas Pipeline Networks." *Economic Theory* 4, pp. 41–65.

Rassenti, S., and V. Smith. 1986. "Electric Utility Deregulation." In *Pricing Electric, Gas and Telecommunication Services*. The Institute for the Study of Regulation, December.

Rassenti, S., V. Smith, and R. Bulfin. 1982. "A Combinatorial Auction Mechanism for Airport Time Slot Allocation." *Bell Journal of Economics*, Autumn, pp. 402–17.

Rassenti, S., V. Smith, and B. Wilson. 2000. "Controlling Market Power and Price Spikes in Electricity Networks: Demand-side Bidding." Economic Science Laboratory Working Paper.

Reynolds, S., 1990. "Cost Sharing and Competition Among Daily Newspapers." Department of Economics, University of Arizona, October.

Smith, V. 1962. "An Experimental Study of Competitive Market Behavior." *Journal of Political Economy* 70, pp. 111–37.

———. 1982a. "Microeconomic Systems as an Experimental Science." *American Economic Review* 72, pp. 923–55.

———. 1982b. "Markets as Economizers of Information: Experimental Examination of the 'Hayek Hypothesis.'" *Economic Inquiry* 20, pp. 165–79.

———. 1987. "Currents of Competition in Electricity Markets." *Regulation* 2.

———. 1988. "Electric Power Deregulation: Background and Prospects." *Contemporary Policy Issues* 6, pp. 14–24.

———. 1993. "Can Electric Power—A Natural Monopoly—Be Deregulated?" In *Making National Energy Policy*, ed. H. H. Landsberg. Washington, DC: Resources for the Future.

———. 1996. "Regulatory Reform in the Electric Power Industry." *Regulation* 1, pp. 33–46.

Williams, A. 1980. "Computerized Double Auction Markets: Some Initial Experimental Results." *Journal of Business* 53, pp. 235–58.

NOTES

1. We use the term "privatization" to describe generically the process of reform of foreign government command forms of organization of the electric industry. In *all* cases major components of the industry have *not* had their ownership transferred from public to private entities. Reform has focused on the use of decentralized spot and futures markets to provide price signals to improve the short- and longer-term management of the industry. The term "deregulation" applies to electricity reform in the United States, where fifty state and one federal regulatory body have regulated an industry already predominantly owned privately, and, although liberalized, that regulation continues to dominate the industry.

2. Hugh Outhred notes that there is ongoing work in Australia under the NECA code-review process to explore practical implementations of network property rights (see www.neca.com.au).

3. Maintaining voltage to avoid "brownouts" requires generators, or special compensating devices, to provide local reactive power. Since generators can produce either reactive or active power (the latter is energy that does work) in variable proportions, (i) is a source of "externality" only if it is not priced, which is the universal practice inherited from centrally owned or regulated systems. We plan experimental designs to price reactive power as just another commodity.

4. Generator (spinning) reserve can be supplied by a market for standby capacity in addition to the energy market (see Olson, Rassenti, and Smith 2001 for an experimental study of such simultaneous markets). A simple such market (without network complications) is provided when you rent an automobile: if you use it, you buy the gas in a separate energy market; if you do not use it, then it is in standby reserve for contingent use. To maintain transmission reserves, lines are typically constrained to carry much less than their thermal capacities by engineers whose zeal in minimizing the risk of losing a line is not necessarily economical. A standard rule, based on n-1 analysis, is to set the capacity of each line in a network so that if any one line goes out the remaining n-1 lines can carry the peak load; if you want still more security n-2 analysis is applied, and so on. Of course this approach begs the question of what price security. Can catastrophic insurance principles be applied with a variable premium that increases with monitored capacity utilization?

5. The impetus for reform was a drastic reduction in the performance of the NZ economy from 1953 to the late 1970s. New Zealand had the world's third-highest per capita income in 1953 (behind the United States and Canada, but tied with Switzerland), and by 1978 had slipped to twenty-second (less than half the per capita income of Switzerland). See McMillan (1998).

6. As a practical matter, because of the cost of metering and monitoring, network pricing always involves a certain amount of aggregation of subsystems into single nodes or paths. Hence, the above principles are indeed conceptual, and only imperfectly captured in any actual operating system. Moreover, low-voltage distribution systems do not follow the square law loss rule at all well, and losses are inevitably averaged across the high density of users.

7. Based on private correspondence with Hugh Outhred.

8. For a critique of the trends in electricity see Smith (1987, 1996), and for studies of smart computer-assisted markets in gas pipeline networks see McCabe, Rassenti, and Smith (1989a, 1990), and Rassenti, Reynolds, and Smith (1994).

5

Electricity Industry Restructuring: The Alberta Experience

Terry Daniel, Joseph Doucet, and André Plourde

This chapter focuses on the particular version of electricity industry restructuring that has been implemented in Alberta—one of the first jurisdictions in North America to embark on the journey to a more market-based system. In evaluating the concept and merits of electricity industry restructuring, this is a particularly appropriate case to examine. The Alberta market has most of the essential elements, and problems, of electrical systems elsewhere, but it is modest in size and somewhat isolated from many complicating factors that tend to confound analysis of larger, more interconnected systems. As such it is an ideal test case for policy analysis. In addition, there are several aspects of the implementation process that have been chosen for Alberta that are unique, interesting, and merit consideration.

INTRODUCTION

Over the past several years, "deregulation" and "restructuring" have, without doubt, entered the popular lexicon. While many industries have gone through the process of transferring decision making and price setting from the confines of regulatory offices to the private market place, in no industry has this process attracted as much public attention as in the electricity sector. The primary reason for this, of course, is that unlike most previous situations, and contrary to expectations, in some electricity markets prices have risen dramatically after the introduction of restructuring. The factors driving this negative turn of events have been many, ranging from the physical nature of electricity networks and the structure of the industry to flawed implementation, poor timing, and just plain bad luck.

The chapter begins with a summary of the development and status of the electricity market in Alberta prior to restructuring. Following this is an account of the motivation for restructuring and the initial steps taken to modify the regulatory control on the industry. The core of the chapter focuses on the particular process chosen to move the province's electrical power system into an almost fully restructured and competitive environment between 1996 and 2002. Issues that appear to be critical determinants of the eventual success of the process in Alberta, and that also shed light on restructuring initiatives in other jurisdictions, are examined.

THE BACKGROUND SETTING: ALBERTA'S ELECTRICITY INDUSTRY

Alberta is somewhat unusual in Canada in that there has never existed a single, vertically integrated Crown (that is, government-owned) monopoly serving the electricity needs of the province. In fact, the historical industry structure in Alberta more closely resembles the U.S. model with several vertically integrated firms operating as franchise monopolies, under cost-of-service regulation, with an integrated transmission network. Interestingly, there has been for some time a relatively high degree of coordination at the provincial level. Beginning in the late 1970s, the three major provincial utilities began dispatching their generation capacity as a single integrated system and were regulated as such by provincial regulatory agencies. Generators continued to operate under a cost-of-service framework. In 1982, reacting to widening differences in generation and transmission costs between franchise areas, the provincial government enacted the Electric Energy Marketing Act (EEMA). Under this act, a provincial agency purchased energy from the "generators" at cost-sensitive prices (basically in a cost-of-service framework) and resold this energy to "retailers" throughout the province at a uniform price. Any remaining price differences between franchise areas were thus limited to differences in the distribution costs, which remained regulated.

Restructuring of the electricity industry was first broached in Alberta in the early 1990s. At this time, the Alberta electricity industry had, at its core, three vertically integrated utilities, accounting for approximately 90 percent of the province's total generation capacity of about 8600 megawatts (MW). Two of these utilities were (and remain) investor-owned: Alberta Power (part of the ATCO group of companies and hereafter, ATCO) and TransAlta Utilities (hereafter, TransAlta). The third major supplier, Edmonton Power (hereafter, EPCOR), was (and remains) owned by the city of Edmonton. TransAlta was by far the biggest player, accounting for more than 50 percent of provincial generation capacity. It also supplied the municipal distribution utilities of Calgary and a number of smaller municipalities, as well as its own franchise area. ATCO owned slightly less than 20 percent of total provincial generation capacity, and EPCOR slightly more than 20 percent. All three were franchise monopolies, regulated by Alberta's Energy and Utilities Board (EUB) (the province's energy regulator).[1] A small municipal utility (owned by the city of Medicine Hat), nonutility generators, and other small power producers provided the remaining 10 percent of generation capacity.

All in all, the three regulated utilities owned approximately fifty generating units, with about 75 percent of this capacity being coal-based. The remainder was about evenly split between natural gas–fired units and hydro. This fuel mix remains an accurate description of utility-owned generation today, since the major players (or anyone else, for that matter) have yet to make significant additions to capacity since the onset of restructuring in January 1996. Table 5.1 shows the merit order of supply in the province immediately prior to restructuring (based on natural gas prices at that time).

Because the utility-owned generation capacity was centrally dispatched starting in the late 1970s, the merit order of supply was public knowledge. This information, as well as the history of operation of the province's utility generation units, meant that the introduction of a competitive bidding process for energy supply in the wholesale market was "relatively" straightforward. Note that table 5.1 does not include any explicit cost information. Under the cost-of-service framework that existed, units were dispatched in order to meet the load, and tariffs were set in order to attain the regulated rate of return. So, the units were dispatched "efficiently," according to underlying economic data, but not as a result of a bidding process.

Finally, it is important to emphasize the extent to which Alberta was and remains an "electricity island." The interconnections with British Columbia through the BC Hydro system to the west, and with Saskatchewan through SaskPower's grid to the east, are relatively small compared to the total capacity and load in Alberta. By the end of 1995, these two lines provided Alberta with access to about 550 MW, roughly equal to 6 percent of the province's total generation capacity.[2] There still does not exist a direct interconnection between Alberta's electricity grid and any U.S. system.

THE INITIAL PERIOD OF RESTRUCTURING: 1996 TO 2000

The Electric Utilities Act of 1995[3] was the first formal step in the process of restructuring of the electric power industry in Alberta. Among other changes

Table 5.1.   Merit Order of Supply in Alberta Prior to Restructuring

Plant	Fuel Type	Capacity (MW)	Retirement Date	Owner
Genesee 1	Coal	385.6	2020	EPCOR
Genesee 2	Coal	385.6	2020	EPCOR
Keephills 2	Coal	382.6	2020	TransAlta
Keephills 1	Coal	382.6	2020	TransAlta
Sundance 6	Coal	365.7	2020	TransAlta
Sundance 5	Coal	354.6	2020	TransAlta
Wabamun 4	Coal	279.7	2003	TransAlta
Sundance 1	Coal	279.7	2017	TransAlta
Sundance 2	Coal	279.7	2017	TransAlta
Sundance 3	Coal	354.6	2020	TransAlta
Sundance 4	Coal	354.6	2020	TransAlta
Wabamun 3	Coal	139.9	2003	TransAlta
Wabamun 1	Coal	64.0	2003	TransAlta
Wabamun 2	Coal	64.0	2003	TransAlta
Sheerness 1	Coal	379.6	2020	ATCO
Battle River 5	Coal	369.6	2020	ATCO
Sheerness 2	Coal	379.6	2020	ATCO
Battle River 4	Coal	147.8	2013	ATCO
Battle River 3	Coal	147.8	2013	ATCO
Milner coal	Coal	109.5	2012	ATCO
Milner gas	Gas	35.4	2012	ATCO
City of Medicine Hat	Gas	50.0	–	–
BC Hydro	(import)	400.0	–	–
Clover Bar 4	Gas	157.9	2010	EPCOR
Clover Bar 3	Gas	157.9	2010	EPCOR
Clover Bar 2	Gas	157.9	2010	EPCOR
Clover Bar 1	Gas	157.9	2010	EPCOR
Rossdale 10	Gas	71.1	2003	EPCOR
Rossdale 9	Gas	71.1	2003	EPCOR
Rossdale 8	Gas	67.0	2003	EPCOR
Rainbow 2	Gas	40.0	2005	ATCO
Rainbow 3	Gas	21.5	2005	ATCO
Rainbow 1	Gas	26.0	2005	ATCO
SaskPower	(import)	150.0	–	–
Sturgeon 1	Gas	10.0	2010	ATCO
Sturgeon 2	Gas	8.0	2010	ATCO

to the industry, the act established a power pool (an independent agency) through which all electricity in the province would flow. Bids to supply power into the pool and to purchase power from the pool would determine, on an hourly basis, the wholesale market price of electricity. The underlying intent of the legislation, as with all restructuring initiatives, was to allow a competitive market to begin providing the price signals that would lead to increased market efficiencies. In particular, by deregulating decisions about new generation, the new structure shifted questions relating to the type, timing, and amount of generation additions from regulatory hearings to the market.

The above approach was essentially quite similar to most restructuring initiatives. Alberta's electricity system was unusual, however, in one important aspect, namely the average cost of embedded generation. Alberta's system was very low cost, and new generation units were (and still are) expected to have average costs above the embedded system cost. Unlike most jurisdictions embarking on restructuring, high cost was therefore not a driver in Alberta. Two important drivers were the desire to streamline (and hence to decrease the cost of) the regulatory process and to move to market-based decisions where possible.

With respect to the construction of new base-load capacity, the regulatory process in Alberta was at a crossroads in the early 1990s. The construction of the province's last large coal-fired plant and its addition to EPCOR's rate base had been fairly contentious and acrimonious.[4] Centralized dispatch, uniform retail pricing, and cost-of-service regulation of generators were at odds with a more competitive generation environment. It is unclear and very uncertain if and how any new large-scale capacity additions would have occurred under the previous regulatory regime. This is one reason that the three utilities supported restructuring. The government of Alberta and the business community's favorable view of market-based systems was another significant facilitating condition. Thus, support for restructuring was broad based.

In this low-average-cost system, one major issue to be addressed is the impact of restructuring, and more specifically the impact on prices of marginal-cost pricing in the wholesale market. In contrast to the situation in many U.S. jurisdictions, a characteristic of the Alberta system was that the marginal cost of any new generation would be higher than the average cost of existing capacity. Restructuring initiatives that would move from an average-cost to a marginal-cost approach to pricing would give rise to the possibility of important rent transfers from consumers to low-cost generators. This issue was dealt with in two successive steps in Alberta, namely the implementation of legislated hedges in the wholesale market and, later, the auction of the energy produced from regulated generation capacity.

The *legislated hedges* introduced in the act in effect tied the generation price of electric energy between January 1996 and December 2000 to the cost of service of regulated units. Owners of these regulated units therefore were unaffected by the price level of the power pool.[5] Retail demand served by these units likewise was unaffected by the price level of the power pool.[6] The hedges were designed with two separate, but closely related, objectives in mind. These were to mitigate the exercise of market power by the three large utilities and to protect existing retail customers and regulated generation units from the pool price.

With respect to the first objective, even though the generation market remained highly concentrated and a bidding process determined the wholesale
price of electricity, the three regulated utilities had no incentive to influence
prices because of the hedges. This was an effective response to the market
power issue, at least in the short term. For most of the 1995–2000 period,
wholesale prices did remain very low.

With respect to the second objective, existing regulated units and retail demand were insulated from the changes in the pool price, while new supply
and demand were meant to see the appropriate price signal of the new pool
price. Existing regulated units were built under the assumption that they
would recover their costs, including a regulated rate of return, over their useful life. Since the pool price was (correctly) expected to be low on average in
the early years and to rise over time as demand rose relative to supply, without the legislated hedges, the relatively new regulated plants would have high

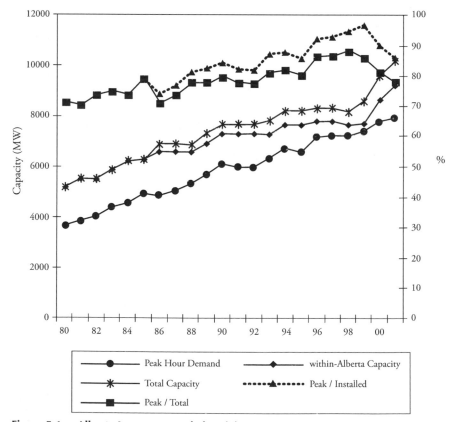

Figure 5.1.   Alberta Interconnected Electricity System

"stranded fixed costs" under low pool prices and the older, depreciated regulated plants would have received a "windfall" gain as pool prices rose. This hedge was calculated so as to give both old and new regulated units a return similar to that expected under the former regulatory regime. Only existing "nonutility" generation, new generation (since it was deregulated, all new generation would be nonutility), and generation quantities above or below rated output faced exposure to the pool price. On the demand side, retailers (again primarily the vertically integrated big three, plus the municipal utilities of Calgary, Red Deer, and Lethbridge) were on the buying side of the legislated hedges, which protected them from volatility in the pool price by giving them a hedged price equal to the average cost of existing utility generation. As a result—and by design—consumers noticed little change from the days of old-style regulation. In fact, the market changed very little with the transition to the pool mechanism in 1996, mostly because a healthy capacity margin kept pool prices low.

Between 1993 and 2000 Alberta's electricity market tightened unexpectedly because of higher-than-forecast growth, both in the Alberta economy and in electricity demand (for example, the latter grew at an average annual rate of about 2.9 percent during that period). During this period entry in the generation segment of the industry was limited to natural gas-fired cogeneration plants tied to large industrial projects.[7] This market tightening is depicted in figure 5.1, which represents the available capacity of the Alberta Interconnected Electricity System at the end of each calendar year. The growth in peak-hour demand outstripped the growth in available capacity (either within Alberta or available through interconnections with BC and Saskatchewan). This is indicated in the three series of figure 5.1 measured against the left-hand axis of the figure. Indeed, the ratio of peak-hour demand to within-Alberta capacity (measured against the right-hand axis in figure 5.1) was some ten percentage points lower in 1993–2000 than in the previous eight-year period, and approximately eight percentage points lower once interprovincial interconnections are taken into consideration.

Interestingly, although the regulatory framework allowed it, no merchant generation plants were built during this period, and this even though the supply cushion was tightening. The fact that pool prices remained relatively low (for the reasons suggested above) is not a sufficient explanation for this lack of investment. A tightening market should have provided the long-term investment signals. However, the level of policy uncertainty that existed during this period, regarding the resolution of the restructuring initiative, effectively froze any capacity expansion plans that firms might have had. Among other things, the uncertainty relating to the continuing monopoly position of distributors over retail customers in their respective franchise area (with the exception of self-generating industrial customers) was a major concern. Since this monopoly position was viewed as temporary, pending the next stage of restructuring, and because the terms of retail competition were not clearly established, there was no one in a position to contract for load growth for that market. Indeed, the only demand to appear as a large identifiable block in the market was industrial load under self-generation that turned to natural gas generation on a project-by-project basis. Potential new generation was thus not able to hedge itself in the retail market, making entry much more risky. In addition, no financial markets for hedging of future supply and demand existed. Given the above, it is not surprising, as will be seen shortly, that EPCOR (the integrated utility owned by the city of Edmonton) and ENMAX (the city of Calgary's distributor and retailer), today's two dominant retailers, were very active in acquiring capacity in the auctions of the regulated energy.

In an effort to create proper price incentives for both supply and demand, and to proceed further along the restructuring path, an amendment to the Electric Utilities Act was passed in 1998. The new legislation called for the removal of the legislated hedges (in January 2001) and for the introduction of new measures to enhance competition among both suppliers and retailers of electricity. For the first time, both the supply and demand sides of the market would be exposed to the market incentives provided by the power pool prices. The primary issues now facing policy makers centered on the means to move toward a competitive market and the speed at which to make this move. There were several important decisions, but the most difficult were on the supply side. At one extreme, a slow path could be chosen that left existing supply in the hands of the current owners but limited the incumbents' participation in new ownership so that over time more widespread competition would emerge. At the other extreme, immediate divestiture by the big three utilities of a large portion of their existing generation capacity could be mandated. Balancing a desire to move rapidly on the restructuring course against a political philosophy that respected private ownership, the government chose an innovative intermediate path.

The chosen plan called for the ownership and operation of existing utility generation units to be left in the hands of existing owners (i.e., no forced divestiture) and for new supply to continue to be totally unregulated. The controversial and somewhat unique new element of the plan obliged the owners of the regulated generation units to sell the ownership rights to the *energy* production from the remaining (regulated) life of these plants at a onetime auction. These ownership rights were referred to as Power Purchase Agreements (PPAs). The purchasers of these PPAs, possibly but not necessarily new players in the market, would then be responsible for bidding the energy into the power pool on a daily basis. In this sense, this onetime auction, with strict limits on purchases, would reduce market concentration of the energy output of regulated units. The hope was that a sizable collection of new participants would be attracted into the supply-side market, creating a new level of competition leading to lower electricity prices. The second goal of the sale of regulated energy was to capture the "stranded benefits" associated with the low-cost embedded generation, as described earlier. The province basically sought to retain the increase in value of these plants (over and above the regulated return) that would be created by wholesale prices moving much higher than what was required under the cost-of-service framework (because of the move to marginal-cost pricing in the wholesale market). The auction of these Power Purchase Agreements was clearly the most adventuresome aspect of the Alberta restructuring process.

Alberta's regulated generation plants, as listed in table 5.1, were grouped into twelve PPAs (see table 5.2 below for a list of the PPAs) for sale.[8] The terms of each PPA contract called for the buyer to pay the owner/operator of the generators (who were, in each case, one of the big three utilities, TransAlta, ATCO, and EPCOR) the marginal generation (primarily fuel) cost of each unit of power produced and sold into the power pool plus a fixed monthly payment representing the annualized, unrecovered capital cost of the plants (so-called capacity payments). The details of these payments, as well as a plethora of contractual parameters, were predetermined by regulatory officials. The buyers of the PPA contracts obtained the right to bid, on a daily basis, the power from the plants into the power pool and to retain the revenues from the resulting electricity sales. The contracts also contained explicit incentives to encourage plant

Table 5.2.   Power Purchase Agreements Offered for Sale in August 2000

PPA	Capacity (MW)	Winning Bidder	Final Price (Millions)	Per-Unit Price (Thousand $ / MW)
Battle River	666	EPCOR	$84.9	127.5
Clover Bar	631	not sold		
Genesee	762	not sold		
Keephills	762	ENMAX	$240.7	315.9
Rainbow	93	Engage	($21.0)	(225.8)
Rossdale	208	Engage	$0.0	0.0
Sheerness	760	not sold		
Sturgeon	18	not sold		
Sundance A	560	TransCanada	$211.9	378.4
Sundance B	706	Enron	$294.8	417.6
Sundance C	710	EPCOR	$268.5	378.2
Wabamun	549	ENMAX	$75.1	136.8
Total Sold	*4254*		*$1,154.9*	*271.5*

operators to strive for operational efficiency and maximum availability of their generation capacity. For instance, energy in excess of the contractual obligation to the PPA owner would belong to the owner of the plant and be available to be bid into the pool.

Most of the PPAs had relatively low fixed charges (reflecting low levels of unrecovered capital) and were expected to sell at high prices reflecting a differential between expected future pool prices and their low marginal energy costs. Some newer plants, on the other hand, had high fixed-capacity charges attached to them, and the associated PPAs were expected to be sold at negative prices in the auction—meaning that the PPA owner would be paid by the province to become the owner of the regulated energy associated with the plant. Overall, the expectation was that the sum of winning bids in the auction would be substantial and would approximate the present value of the difference between the expected pool price and the total costs of generation embedded in the PPAs. This surplus—to be retained by the province and accumulated in a "balancing pool" (a fund associated with the power pool)—would then be returned in some form to consumers. This transfer served as compensation to consumers for having to buy electricity from retailers whose prices, based on the power pool, were expected to reflect the (higher) cost of new generation, rather than the (relatively low) average embedded cost of existing generation (about $30 per megawatt-hour in 1998).[9] In other words, this surplus was meant to reflect the "value" to Alberta consumers of the embedded regulated generation, and the intent was to return these rents to them in a manner that did not compromise market efficiency.

Because the PPA auction process had two distinct objectives, namely to introduce structural relief in the generation/energy supply segment in order to foster a competitive environment and to capture the stranded benefits in existing regulated generation, its eventual success was highly dependent on the level of competition that the auction would create. More competition in the auction would lead to greater value obtained by the province (stranded benefits). Unfortunately, the number of firms that registered for the auction was almost identical to the number of PPAs that were for sale. Most of these firms did place deposits that would have allowed them to bid on more than one of these contracts.[10] In terms of the number of bidders, this was probably close to the minimum that the government would have accepted without withdrawing the auction (though the government criteria on this were never made public). While somewhat below early estimates, and certainly disappointing, these values gave hope that there would be "reasonable" competition in both the auction and the subsequent wholesale market.

The format of the auction was a series of ascending price-bidding rounds— with four to six rounds per day—conducted over a three-week period. Bidding was simultaneously open on all PPAs in each round and was conducted using a secure Internet site. The rules of the auction were such that a firm had to keep actively bidding in order to retain its initial eligibility (as determined by its deposit). For example, if a firm had placed a deposit so as to have the right to bid on 800 MW of power, it had to place active bids in each round for this much capacity. (The rules allowed some latitude in this regard, particularly in the early rounds—the specifics are detailed in Charles River Associates Incorporated and Market Design Inc. [1998].) Early on in the auction, however, most firms appeared to let their eligibility drop to a level such that the amount of power being sold was only slightly higher than the number of eligible bidding units (because the results of each round of the auction have never been made public, this conclusion is open to speculation). As a result, the competition for each PPA was far from fierce. A total of eight PPAs were sold to five separate firms. Table 5.2 lists the PPAs along with the results of the auction.

The PPA contracts associated with two recent base-load plants with high fixed payments (Genesee and Sheerness), which were expected to draw negative bids, did not sell; nor did two gas-fired peaking plants (Clover Bar and Sturgeon). The prices commanded by the PPAs that did sell appeared to be low, reflecting either market anomalies, a flawed auction process, or a combination of these factors. Each of these possibilities will be examined in turn.

The design of auctions, particularly for markets with small (two to twenty) numbers of participants, has been the focus of a considerable amount of research over the past two decades (see Wilson 1993). In particular, the application of game theory to the analysis and design of auctions has produced a large number of important theoretical and empirical results (see Klemperer 1999 for a comprehensive survey of this work). In 1994, this research began to influence practice with the auction of radio spectrum rights in the United States by the Federal Communications Commission (FCC). Several prominent game theorists were invited to propose designs for this multibillion-dollar auction of bandwidth for wireless communications. The FCC chose a design submitted by Paul Milgrom and Robert Wilson of Stanford University, which involved an ascending-bid process whereby blocks of wireless rights across the nation were simultaneously auctioned. The auction generated more than $(US)7 billion in revenues for the FCC and was generally considered to be a success (McAfee and McMillan 1996).

Subsequent analysis (e.g., Weber [1996]), however, has cast some doubt on the efficiency of the process and has suggested that informal collusion (particularly in the case of the "B-Block" auction where the number of bidders was small relative to the number of blocks for sale) may have led to sale prices well below competitive levels. Because the auction mechanism used to sell the electricity PPAs in Alberta was almost identical to that used for the spectrum rights in the United States, similar concerns might be raised about the power auction. One notable change in procedure was to limit information on bidder identity between rounds of the Alberta PPA auction. However, the number of participants was small enough that it is conceivable that implicit cooperation was possible ("don't raise the bid on mine and I won't bid on yours"). This same lack of public information on the identity of the bidders also makes subsequent analysis of bidding behavior difficult.

Recent research points to another potential problem with open ascending-price auctions that may have been a factor in keeping prices low in the PPA auction. Since the early 1970s, considerable attention has been focused on the "winner's curse" issue in common-value auctions. An auction is termed "common-value" if the item being sold has the same economic value to all bidders. In most circumstances, this common value is uncertain—no bidder knows exactly what the item will eventually be worth should he or she succeed in acquiring it. A classic example is an oil lease. At the time of the auction, no one (including the auctioneer) knows the value of the oil beneath the ground, but all bidders have estimates of that value and each knows that whatever the ultimate quantity discovered, the extracted value will be the same in each party's hands. Early empirical research on off-shore oil leases in Louisiana indicated that, in such circumstances, the winning bidder was usually the one with the most (overly) optimistic estimate (seismic survey) of the value of the lease and that subsequent revenues from drilling often failed to match the price paid at auction. The winner of the auction was "cursed" with a loss on the transaction.

Theory and practice over the intervening years have shed light on the correct level of bid discounting required so as to avoid the curse when bidding in a common-value auction. The resulting shaded bids should, on average, produce a competitive level of profits for both the winning bidder and the auctioneer (as a function of the number of bidders—with the auctioneer gaining a larger share of the profits from the transaction as the number of bidders increases). In a recent paper, however, Klemperer (1998) has shown that even a mild asymmetry in the value that individual bidders obtain from the sale or use of the item being auctioned can have an "explosive" (downward) effect on the expected revenue obtained by the auctioneer. If it is common knowledge among the potential bidders that the assets being auctioned will be more valuable in one bidder's hands than when owned by any of the others, then the optimal bids for each party to place in the auction are only a fraction of what they would otherwise be. In other words, the privileged bidder can expect to face greatly reduced competition for the item being sold.

In the Alberta PPA auction, it was common knowledge among bidders (and most industry observers) that some firms, in particular the two large municipal utilities in Edmonton and Calgary, EPCOR and ENMAX, appeared to be positioned to obtain marginally greater value from the acquisition of wholesale power packages than other registered bidders. Both of these firms had an existing (though no longer guaranteed) customer base with its requisite attributes of customer loyalty and billing efficiencies. They were "long" on the demand side of the new power market and the wholesale power parcels (PPAs) nicely complemented their portfolios. Most of the remaining bidders, in the short to medium term at least, faced the prospect of extensive exposure on the supply side if they were successful in obtaining PPAs. Theory predicts that in such a circumstance, these firms will lower their bids significantly even relative to their real competitive disadvantage.

In light of the above observations, the results of the August 2000 PPA auctions will be examined. Each PPA contract is a complex instrument whose value is a function of numerous variables. Many of these variables are highly uncertain over the horizon (in some cases, twenty years) of the contract. The value of a PPA contract is the net present value of the stream of differences between the hourly pool price for electricity in Alberta and the marginal cost of generation (primarily the fuel cost of gas or coal) minus the fixed monthly capacity payments as specified in the contracts. Table 5.3 illustrates some approximate calculations for one of the large coal PPAs. Shown are the internal rates of return for a variety of average pool prices over the life of the contract given the actual auction sale price. Immediately obvious is the extreme sensitivity of the value of these contracts to future electricity prices. At prices that prevailed in the period prior to 1999 (less than $35 per MWh), the contracts are only marginally profitable. At prices that have prevailed since the auction (more than $60 per MWh), they are extremely lucrative!

With a pool price of $76 per MWh, for example, the payback period for the purchase of such a PPA is just one year. Although not shown in table 5.3, similar results are obtained for other PPA contracts. Based on this cursory analysis, the winning bids appear to have been quite low compared to the realized

Table 5.3.   Rates of Return Implied by Auction Sales Price for Various Pool Prices

Average Pool Price	Internal Rate of Return
$30/MWh	2%
$40/MWh	27%
$50/MWh	48%
$60/MWh	68%
$70/MWh	87%

Assumptions: The rates of return are calculated on the basis of the "Sundance A" PPA with an assumed coal price of $4.50 per MWh and the indicated pool prices (in Canadian dollars at 1999 prices) over the twenty-year life of the plant, an auction sale price of $211.9 million, and an annual capacity payment of $86.5 million.

value of the contracts. Some possible explanations for the low prices bid in the auction are:

- Bidders adopted very conservative forecasts of future pool prices (well below the prices that had prevailed over the past year in Alberta);
- Bidders discounted future profits heavily in light of uncertainties over the long horizon of most of the contracts; or
- Less-than-competitive conditions prevailed in the auction for reasons such as those described previously in this section of the chapter.

Not surprisingly, the fact that the balancing pool netted a total of just over $1 billion from the auction (see table 5.2), well short of initial expectations, proved to be quite controversial, especially as wholesale prices surged in late 2000. As explained earlier, these funds were to be used to compensate consumers for the higher cost of electricity under the new marginal-cost pricing scheme of the competitive generation market. The combination of lower than expected auction revenues and higher than expected pool prices effectively meant that rebates to customers in the year 2001 exhausted the balancing pool funds. Fortunately, pool prices exhibited a downward trend in the eighteen-month period after January 2001, thus reducing the public pressure to continue customer rebates.

On the more important (in the long term at least) issue of increased competition on the supply side of the market, the number of firms bidding power into the pool has now more than doubled. The jury remains out, however, as to whether this will be sufficient to support a competitive market. The key will likely be the rate and ownership pattern of new supply. A recent report issued by the Alberta Department of Energy paints an optimistic picture with a forecast of 4800 MW of new supply coming on line by 2007. There is some evidence, however, that the incentives for non-price-taking behavior remain high. In the period after January 1, 2001, the Market Surveillance Administrator (MSA) of Alberta has issued a sequence of reports in response to claims of market manipulation by some suppliers. While enumerating several structural factors behind the prices, including high natural gas prices and the influence of shortages in California and rising demand (as evident in figure 5.1), the reports have also discussed "behavioral" factors. Questions were raised about bidding and supply withholding on the supply side. To date, the MSA has concluded that while the potential for the exercise of market power remains a concern in Alberta, data and evidence are lacking to draw firm conclusions that observed behavior is either inappropriate or worthy of additional regulation. In light of similar behavior in the first years of restructuring in Britain (see

Green and Newberry 1992; Newberry 1998), it would not be surprising if attempts to exercise market power occurred in Alberta as well. We return to these questions below.

On the retail side of the market, initial plans were to move to open competition on January 1, 2001. At that time, all retail customers were expected to have freedom of choice with respect to their supplier. Prices would be determined through negotiated contracts and were expected to range from spot-market pricing to long-term arrangements. The retail market would be opened up to anyone who wanted to participate. To provide residential and small commercial customers with a period of grace in order to understand the new market and to adapt to the potential volatility of prices, a Stable Rate Option (SRO) was included in the legislation (the name has since been changed to Regulated Rate Option [RRO]). These smaller consumers of electricity were given the option of continuing to buy power from their current franchise utility, under terms subject to regulatory oversight by the EUB, for up to five years. The existing utility was required to supply this contract, while its customers surveyed the new marketplace and sought retail contracts to suit their needs. The distribution network, however, was to remain a regulated segment of the industry, populated by the three regulated utilities and a few wire-owning municipalities. After January 2006, all retail franchise rights and obligations were to disappear from the Alberta electricity market.

In November 2000, in the face of skyrocketing wholesale prices, the government of the province placed a hold on the plans for a deregulated retail market. Retail prices were capped for 2001, effectively stalling the restructuring of the demand side of the market. At the same time, financial contracts for the years 2001 to 2003 on the energy (both base and peak load) from the PPAs that did not sell in the original auction in August were sold at a second auction. The balancing pool retained the dispatch rights to this energy (which it continues to bid into the pool), but the contracts provided fixed-price hedges for those years for both the balancing pool and a large number of major consumers and brokers of energy who purchased them at the auction. The prices obtained ranged from $117 per MWh for base-load power in 2001 to $153 per MWh for peak periods with an average of $122 per MWh. For 2002, the average price was $67 per MWh and for 2003 it was $60 per MWh—providing a consensus forecast of future electricity prices in the province.

The average price obtained on the 2001 blocks of base-load energy sold at this second auction, approximately 11 cents per KWh, was specified to be the regulated rate for 2001 for small consumers (residential and small commercial), and the nearly $2 billion profits from the two auctions were used to provide subsidies to keep residential and commercial electricity bills near their year 2000 levels.

With the reelection of the provincial Conservative government—the architects of the electricity-restructuring program—in a provincial election in March of 2001, a mandate of sorts was provided to continue the process. In light of the high-profile problems in California, the initial questions remained concerning restructuring: How far and how fast?

THE NEXT PHASE: JANUARY 2001–MAY 2002

In the court of public opinion, restructuring in any industry is evaluated primarily on the impact on consumer prices. Figure 5.2 shows the pattern of electricity prices in Alberta over the period since the auction of wholesale power to May 2002.

While the next phase of restructuring did not come into effect until January 2001, the extremely high prices in the fall of 2000, immediately subsequent to the auction, were often attributed to the process of restructuring. An important issue in the provincial election of March 2001 was electricity

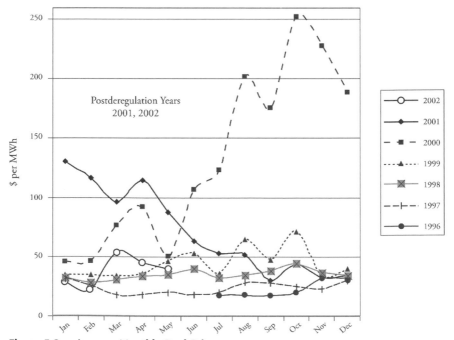

Figure 5.2.   Average Monthly Pool Prices

restructuring and its impact on consumer prices. With hindsight, the above price pattern can be explained by several observations, most unrelated to restructuring in Alberta. The Alberta market, as previously noted, is small and isolated. When all generating plants are operating, the supply/demand balance is tight but adequate. A single base-load plant of 300 MW size, however, represents about 5 percent of average daily demand and about one-half the typical operating reserve maintained in the province. In August 2000, at almost the precise time of the PPA auction, one such plant failed and remained out of service until May 2001. This failure made the Alberta system very sensitive to normal day-to-day fluctuations in supply or demand. Pool prices were often driven to high levels when any other plant went off-line. At the same time, the power crisis in California had an important impact on Alberta prices. In previous years, shortfalls in Alberta supply were often met with increased imports over the tie-line to British Columbia (whose capacity was approximately that of a large base-load plant). With extremely high prices and shortfalls on the Pacific coast, BC Hydro bid and offer prices into the Alberta pool skyrocketed, and the net flows on the tie-line reversed, with Alberta becoming a net exporter rather than importer of power. Table 5.4 illustrates this changing pattern.

In November of 2000 a modification to the wholesale pricing system was instituted in Alberta whereby imports and exports were no longer able to set the market-clearing price in the pool (though they were still dispatched in the merit-order curve). This explains the drop in pool prices after October. Also of importance is the fact that Alberta, which is winter peaking, had an extremely mild winter in 2000–2001. In addition, as figure 5.1 reminds us, the installation of new capacity was expanding supply: about 1600 MW of additional within-Alberta capacity was installed between the end of 1999 and December 2001—an increase of almost 15 percent.

The third important factor in the high prices of 2000–2001 was the high price of natural gas. Peaking power in Alberta is gas-fired, and the record high gas prices of the winter of 2000–2001 were reflected in high peak electricity prices. On a typical day, the marginal plants in Alberta between

Table 5.4.   Changing Import/Export Pattern

Month	Average Daily Peak Exports (MWh)	Average Daily Peak Imports (MWh)
Jan 00	75	175
Apr 00	100	200
Oct 00	325	350
Jan 01	500	100
Apr 01	500	75

11:00 A.M. and 8:00 P.M. are gas plants with a combined capacity of approximately 1000 MW. The Alberta balancing pool controls the bidding of
about one-half of this power into the pool (the plants that did not sell at the
PPA auction). Beginning in January 2001, the balancing pool announced
publicly that it would bid this power in at marginal cost, reflecting the
price of a natural gas hedge that had been purchased to partially shield
against the extreme prices of the winter. Even so, the offer price of this
electricity was about $97 per MWh. In June, with decreasing gas prices,
the bidding policy of the balancing pool was changed and this power was
offered into the pool at a price reflecting the current spot prices of natural
gas. Figure 5.3 shows the dramatic effect of this change of policy, combined with decreasing gas prices and the return to service in May of the
base-load power plant that had been sidelined since August. Shown is the
merit-offer curve for the third Wednesdays of January and July 2001. In
July, this marginal peaking electricity was being offered at $42 per
MWh—less than one-half the price of the winter months. This plateau in
the offer curve is often the one that sets peak hour prices in Alberta. In addition, power from the cogeneration plants in the province is also now being bid into the pool at lower prices reflecting both reduced gas prices and,
more importantly, the need to come in below the prices of the gas peakers.
Clearly the price of natural gas has a significant impact on the price of
electricity in Alberta.

On the question of whether collusive or "uncompetitive" behavior has
been a factor in higher prices since restructuring, the jury is still out, although increasingly such behavior appears unlikely to have been a factor.
The factors mentioned, plant failure and export and gas prices, explain most
of the fluctuations. With respect to the base-load plant that failed, the oper-

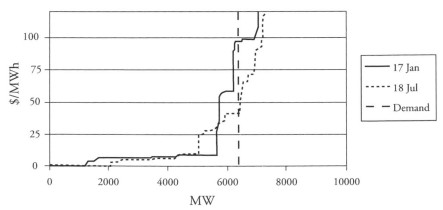

Figure 5.3.   Peak Period Supply, 2001

ator of that plant appealed to the regulatory authority to have the outage declared force majeure — beyond their control. Under such a ruling, the operator would have been able to recoup most of its losses for lost output, at the expense of consumers. (Under the incentive clauses built into the PPA agreements, if a plant is down for "controllable" reasons, the operator must reimburse the holder of the PPA for that plant for the lost output.) In a somewhat surprising ruling, the regulatory authority ruled that the outage was not managed properly and held the operator responsible for the lost output. This ruling should further ease public concerns that plant operators and power wholesalers can readily manipulate pool prices and extract excessive rents in the deregulated environment.

THE PATH FORWARD

At this point it would be difficult to predict how Alberta pool prices will evolve in the short run. The following elements will no doubt have considerable influence on the prices that will prevail. Natural gas prices will of course be key, because in the Alberta system gas is the marginal fuel a great deal of the time. Because the Alberta economy continues to be quite strong, it is anticipated that demand growth will be strong. Just how strong demand growth is will obviously have an influence on prices. Normally, relatively high prices would have some dampening influence on demand. In fact, pool authorities did note a measurable decrease in demand over the 2001–2002 winter that was at least partially attributable to high wholesale prices. Precipitation in the Pacific Northwest, coupled with continuing market tightness in California, will also play a role in Alberta's market. Even though imports and exports do not set the Alberta pool price, the shortage of power in the West will affect market dynamics in Alberta. The internal supply situation in Alberta should be moderately better in 2002–2003. A small amount of new capacity is scheduled to come online before December 2002. Of course the supply margin will continue to be tight, meaning that any prolonged forced outage at any base-load plant could potentially cause important price increases in the pool.

Several longer-term issues need to be addressed in Alberta. Of most immediate concern is the question of construction of new merchant capacity in the province. A large number of projects were announced in 2001. Interestingly, three of these are large, coal-fired base-load units. Because of the long lead times necessary, these large units were not expected to provide relief to the market until 2004–2005 at the earliest. In order for new capacity to be built quickly, the regulatory framework has to be clarified. This includes the permitting process and the regulations and policies in the new market. Regarding the latter question, many different customer classes in Alberta received consumption rebates in 2001 and were thus shielded from the pool price. The impact of these rebates on the development of retail competition is difficult to ascertain. Having said this, the future of retail competition, especially for the residential and commercial segments, is still uncertain.

Also related to clarifying policy, the rapid resolution of the regulated capacity that was not sold in the PPA auctions is necessary. As of the summer 2002, the balancing pool controlled (dispatched) close to 2000 MW. From time to time, depending on economic conditions, the pool may dispatch some of this capacity in an uneconomic fashion, thus distorting the pool price. This obviously can affect the value to other pool suppliers. Importantly, it distorts the signal to potential entrants in the generation market. Getting the pool administration out of the operation of generation units would most certainly benefit the development of the market.

Finally, the electricity industry needs to continue its development along several related directions. A nascent electricity forward market does exist in Alberta, but it is still immature and very thin. The availability of risk-hedging opportunities will be important for continued development of merchant capacity, as mentioned earlier. Along the same lines, more demand-side responsiveness, possibly through better use of retail metering, would also benefit the development of the market.

REFERENCES

Alberta Department of Energy. www.resdev.gov.ab.ca/electric/rgeneral/keynumbers.htm.

Charles River Associates Incorporated and Market Design Inc. 1998. "White Paper on PPA Auction Design Issues." December.

Green, Richard J., and David M. Newbery. 1992. "Competition in the British Electricity Spot Market." *Journal of Political Economy* 100 (5): pp. 929–53.

Henwood Energy Services Inc. 1999. "Alberta Electricity Market Assessment." Report prepared for the Independent Assessment Team, Alberta. Sacramento, CA, July.

Klemperer, Paul. 1998. "Auctions with Almost Common Values." *European Economic Review* 42 (3–5): pp. 757–69.

———. 1999. "Auction Theory: A Guide to the Literature." *Journal of Economic Surveys* 13 (3): pp. 227–86.

London Economics Inc. 1998. "Options for Market Power Mitigation in the Alberta Power Pool." Report prepared for the Alberta Department of Energy. Cambridge, MA. January.

McAfee, R. Preston, and John McMillan. 1996. "Analyzing the Airwaves Auction." *Journal of Economic Perspectives* 10 (Winter): pp. 159–76.

Newberry, David M. 1998. "Competition, Contracts, and Entry in the Electricity Spot Market." *Rand Journal of Economics* 29 (Winter): pp. 726–49.

Weber, Robert J. 1996. "Making More from Less: Strategic Demand Reduction in the FCC Spectrum Auctions." Kellogg Graduate School of Management, Northwestern University.

Wilson, Robert B. 1993. "Design of Efficient Trading Procedures." In *The Double Auction Market: Institutions, Theories, and Evidence*, ed. D. Friedman and J. Rust, pp. 125–52; Reading, MA: Addison-Wesley Publishing Company.

NOTES

1. The utilities were regulated under a traditional cost-of-service framework, with prices set to achieve a regulated rate of return.

2. The 550 MW was made up of 150 MW to Saskatchewan to the east and 400 MW to British Columbia to the west. The British Columbia link is now 800 MW.

3. This act of the provincial legislature came into effect in January of 1996.

4. The two units of EPCOR's Genesee station were commissioned in 1989 and 1994, respectively.

5. All utility units built before 1995 are referred to as "regulated" units.

6. The pool price was the price seen by new merchant capacity built after 1995 and new retail demand.

7. As in other jurisdictions in North America, natural gas was considered to be the fuel of choice for new electricity generation (because of improvements in natural gas generation technologies, relatively low natural gas prices, and relatively short construction lead times, among other reasons).

8. Note that the regulated hydro plants were not included in the auction for reasons related to strategic operation of hydro in the pool. The operational control of hydro is to be maintained by its original owner, TransAlta, with some contractual oversight by the power pool.

9. Unless otherwise specified, all dollar values are expressed in units of Canadian currency. At exchange rates prevailing in 2002, $(C) 1.00 = $(US) 0.65.

10. The seven firms registered for the auction placed deposits, in total, that would have allowed them to bid on 10,000 MW of power. The total capacity of the PPAs being sold was just over 6,400 MW.

6

Transactions Costs and the Organization of Coordination Activities in Power Markets

Craig S. Pirrong

Since Adam Smith described the operation of the "invisible hand," economists have recognized that decentralized trading activity can efficiently coordinate complex resource allocations. Thanks to Coase (1937, 1960), however, economists now also recognize that decentralization is never complete. Within firms and other formal organizations, "visible hands" coordinate the allocation of the resources.

Coase's seminal work initiated a major research program dedicated to investigating the boundaries between centralized ("firm") and decentralized ("market") resource allocation mechanisms. Two outgrowths of Coase's work, transactions-cost economics and property-rights economics, have helped improve our understanding of where these boundaries are drawn.

This work is particularly relevant to the electrical power industry because this industry is now undergoing a massive restructuring in the United States, Europe, Canada, South America, and Australasia. Redrawing boundaries is a crucial element of this process. Historically, the electrical power industry has been highly centralized, with virtually all important resource allocation decisions delegated to vertically integrated firms or government agencies. Heretofore, decentralized market mechanisms played little role in power markets. All this is changing, however. Restructuring efforts around the world differ in the particulars, but share one common feature. All are attempting to facilitate the development of decentralized trading mechanisms for the purchase and sale of electricity. Decentralization is accomplished in part by "dis-integration," that is, the organizational separation of traditionally integrated generation, transmission, and distribution functions. These efforts are predicated on the belief that markets can better coordinate energy transactions. In the restructured setting, coordination of the use of the transmission system is delegated to a central agent, such as an Independent System Operator (ISO) or an investor-owned (or state-owned) transmission company.

This raises a crucial question: How far can decentralization go in power markets? This chapter attempts to analyze the costs and benefits of decentralization in power markets using the tools of transactions-cost and property-rights economics. The basic conclusion is that complete decentralization is infeasible in power markets owing to the physical characteristics of power transmission systems. In particular, reliable operation of a power system is a public good, which creates well-known collective-action problems that can be mitigated through the creation of a central coordinating agent. Moreover, the time scale of power system operations can create acute holdup problems in decentralized contracting setups. Efficient mitigation of these problems implies that the allocation of control rights can have an important impact on the efficiency of the system.

Transactions-cost and property-rights considerations also imply that the form of organization of the coordinating agent and the legal environment in which it operates have important efficiency implications. The traditional model in which a vertically integrated utility generates and transmits power within a monopoly service territory mitigates holdup problems and facilitates the coordinated use of the transmission system, but may result in market power in generation because a single firm has the franchise to serve customers within its territory. Dis-integration facilitates competition in generation, but requires the creation of some organization to coordinate use of the transmission system to maintain system reliability. Due to the complexity of power systems, independent agents responsible for this coordination may have substantial information advantages over their customers; in particular, they have considerably more knowledge about the physical characteristics of the grid. Complexity may also make it impossible to contract on reliability or to use reputational mechanisms to motivate the coordinator to act efficiently. Under these circumstances, "low-powered" incentive regimes (such as profit regulation of transmission companies or the formation of nonprofit coordinating agents in which there is an attenuated relation between agents' compensation and their performance) are likely to be efficient.

Moreover, the governance structure of the coordinator matters. Governance structure specifies who owns the relevant assets, who makes decisions, and the decision-making process. Governance structure affects rent seeking and negotiation costs. In a dis-integrated power market, a nonprofit organization controlled by stakeholders who consume, produce, or market power (such as an ISO) is plausibly a transactions-cost-economizing form of organization. The characteristics of dis-integrated power markets (where heterogeneous agents produce a joint product) are similar to those observed on financial exchanges such as the New York Stock Exchange and the Chicago Board of Trade. Financial exchanges have in fact adopted the nonprofit cooperative stakeholder controlled organizational form and thus may serve as a role model for power markets.

Most who study or work in power markets take the need for central coordination of crucial reliability-related functions for granted. There is much dispute, however, over the most efficient way to perform and organize this coordination function. This chapter endeavors to go back to first principals to identify the basic economic considerations that preclude complete reliance on coordination through market transactions. An understanding of the fundamental transactional characteristics that necessitate organization and formal governance informs the debate over the most efficient institutional framework.

TRANSACTIONS COSTS AND THE NEED FOR CENTRALIZED COORDINATION IN POWER MARKETS

The generation, consumption, and transmission of electricity make up an integrated physical system. Generators produce power and inject it into the transmission grid; loads consume the power. Efficient operation of the power grid requires the system to be balanced at all times, with load equal to generation plus transmission losses.[1]

This physical system is subject to a variety of constraints. Generators operate subject to capacity constraints and sometimes to energy input constraints.[2] They may also be constrained by regulatory limits on their output of pollutants.[3] Generators also face constraints on their ability to adjust output (so-called ramping constraints). Transmission is constrained as well. Transmission lines have thermal limits, stability constraints, and voltage constraints.

The various constraints in the power system are time varying and random. Generators must shut down for maintenance periodically. Random equipment failures ("forced outages") can force generators to shut down unexpectedly. Generator efficiency and transmission line capacity vary with atmospheric conditions. Transmission lines can fail (due to storm damage, for instance). Moreover, power does not flow along any prespecified "contract path" between buyer and seller. Instead, Kirchoff's laws imply that power is subject to "loop flows." That is, power flows through the transmission grid via the path of least impedance, determined by the geographic pattern of generation and load rather than via a route specified by the producer and consumer. As an example of loop flows, in 1989 power generated in Ontario and sold to New York flowed over power lines in Michigan, Ohio, Kentucky, West Virginia, Virginia,

Maryland, and New Jersey as well as along the direct connection between Ontario and New York (Centolella 1996). The buyer and seller had no influence over these physical flows. Together these factors in turn imply that constraints on transmission depend upon the entire configuration of the grid—where power is produced, where it is consumed, and the transmission lines connecting producers and consumers (Centolella 1996). Thus, random fluctuations in spatial load patterns affect transmission capacities.

Violation of the constraints on a power system can cause disruption of service to large numbers of power consumers. Consumers may suffer voltage reductions (brownouts) or complete power loss (blackouts). These disruptions can be extremely costly. Losses from disruption range from the minor (spoiled food in the refrigerator, damage to electrical appliances, missing the latest episode of a favorite television show) to the catastrophic (mass looting *à la* New York in 1965, destruction of generators, deaths due to loss of essential services).

The integrated nature of the network implies that efficient operation and the avoidance of service disruptions requires coordination among network participants. That is, efficient response to an event such as the forced outage of a generator or a transmission line, or a spike in load, may entail adjustment of production at multiple other plants to maintain safe operation of the grid. Moreover, in many cases these adjustments must occur over very short time frames measured in minutes.

Of course coordination problems are not unique to the power market. Indeed, all economic problems are at root coordination problems. Any economic system must coordinate activities of disparate actors to achieve efficient outcomes. However, the conditions in the electricity market pose acute challenges to coordination; the number of agents whose actions must be coordinated is large, these agents are scattered over wide geographic regions (spanning entire regions of a country such as the United States), their actions must be coordinated continuously, and the consequences of coordination failure are huge.

The Coase theorem implies that if transactions costs are zero, any coordination problem can be solved efficiently through the allocation of property rights. In the absence of transactions costs, holders of property rights will exchange these rights to ensure that resources flow to their highest value use. That is, trading of property rights coordinates the activities of individual economic actors to ensure an efficient outcome.

The Coase theorem is as applicable to the power market as to any other. Consider, for example, the coordination of generators and consumers of electricity in a power network in the absence of transactions costs under some set of property rights specifying the ability of generators to inject power into the network and the ability of consumers to draw power from the network. For instance, assume that generator A has the right to inject X megawatts of power at node α of the network. Under some possible configurations of generation availability, load, and transmission, it may be the case that A would impose large costs on others connected to the network if it exercised this right. As an example, by taking this action when a particular transmission line is out of service, the generator could cause another line to fail, leading to a blackout in a large urban area. In the absence of transactions costs, those adversely affected would negotiate a transaction with A whereby he would refrain from undertaking this destructive action. This transaction would make everyone better off. The ability to transact in this manner would implement the coordinated responses required to ensure efficient operation of the network at all times. Thus, in the absence of transactions costs, decentralized actions of economic agents are sufficient to ensure efficient coordination. No central authority is required.

This analysis implies that any need for a centralized agent with the authority to direct resource allocation stems from the existence of transactions costs. The source and severity of transactions costs depends on the particular economic context.

In the power market, transactions costs arise from at least two sources. It is clear, moreover, that these transactions costs are potentially large enough to make complete decentralization inefficient.

The first source of transactions costs is collective-action problems. Failure to achieve an efficient coordinated response to a demand or supply disruption in the power market usually affects a large number of individuals. For instance, a blackout imposes costs on all residents of the affected area. The affected area may be quite large, encompassing entire states. Thus, a coordination failure is effectively a public bad that affects a group of individuals simultaneously. Equivalently, efficient coordination is effectively a public good. It is well known that provision of public goods is affected by collective-action problems—problems that arise from the difficulty of getting many self-interested individuals to act simultaneously in the interest of efficiency. These include the incentive to misrepresent willingness to pay and free riding (Fudenberg and Tirole 1992). Each individual is likely to reason that (a) the quantity of the public good provided will not depend strongly on the price that he pays for it (since any individual pays only a small fraction of the total cost), but that (b) the price he is charged is likely to depend on how much he reveals that the good is worth to him. Under these circumstances, each individual has a strong incentive to say that he is willing to pay only a small amount for the good even if it is very valuable to him. Low-balling is unlikely to affect the amount of public good provided (because this depends on the responses of all potential users, not a single individual's response) but may reduce the amount that the individual pays for the good. If everyone does this, however, revenues from the sale of the public good will be inadequate to pay for its provision. Thus, even if everyone values the good highly and would prefer that a large quantity of it is supplied, the amount actually supplied will be inefficiently small.

In addition, coordination is subject to scale economies. Coordination requires monitoring of the network. Monitoring by multiple agents would be duplicative and wasteful. Moreover, efficient monitoring requires investment in special skills and information. Acquisition of these skills and information by all those hooked up to the grid is also duplicative and wasteful.

The collective-action and scale-economy problems may be addressed through the creation of a firm or organization that acts as an agent for consumers of power. This "solution" raises as many questions as it answers. How does this agent (who I shall refer to as hereafter as the Agent or the coordinating Agent) ascertain consumers' willingness to pay for reliability? Moreover, even if the Agent has perfect information about consumers' willingness to pay, will he act in their interest? That is, how can the Agent be motivated to undertake any and all welfare-enhancing transactions? These questions are central, and are addressed in some detail below.

There is another source of transactions costs even if there is an institutional design that addresses the information and incentive problems facing the Agent. In particular, negotiation costs—a type of transaction cost—are likely to be large. Negotiation costs will exist—and are plausibly large—in power markets because of what are sometimes referred to as temporal specificities.[4]

Temporal specificities may exist when agents must act quickly to maximize gains from trade. The necessity for rapid action means that it is difficult—and perhaps impossible—to find alternative trading partners if one agent refuses to participate in the wealth-maximizing action. In these circumstances, an agent can credibly threaten not to take the wealth-maximizing action unless he receives the lion's share of the gains from trade. This is referred to as a "holdup" in the transactions-cost literature. Other agents may acquiesce to this extortion because if they do not (a) they have no time to find a trading partner to take the extortionist's place and (b) there may be no gains from trade at all. That is, the difficulty of finding alternative trading partners in a hurry can give each agent considerable bargaining power.

It is likely that temporal specificities are acute in power markets. Consider the Agent acting on behalf of power consumers when a transmission line fails. The Agent needs to negotiate a transaction (or a series of transactions) with owners of generation to forestall a catastrophic system failure. Assume that the efficient response to the line failure requires generator *G1* to curtail output and generator *G2* to increase output. These adjustments must be made in minutes. If this efficient response is not implemented, the second-best response may result in large costs (due to a blackout, or to the necessity of operating generator *G3*, which is much more expensive than *G2*.)

Large quasi-rents will exist under these circumstances.[5] That is, there is a large gap between the cost of the efficient action (which is roughly equal to the difference in marginal costs between *G2* and *G1*) and the benefit thereof (which may be immense if failure to implement the efficient response leads to a blackout). The quasi-rents are likely to be particularly large because action is required on very short notice—in minutes. The short response time sharply limits the alternative actions available to mitigate the effects of the transmission line failure. In this case, temporal specificities are acute.[6]

The crucial insight of transactions-cost economics is that quasi-rents resulting from specificities create transactions costs.[7] Parties have incentive to expend resources to capture the lion's share of the quasi-rents. If the quasi-rents are large, these rent-seeking expenditures—real resources used to gain a wealth transfer rather than to create wealth—will be large as well. Rent-seeking expenditures are a pure waste. Moreover, if the bargaining parties are differentially informed (as is likely to be the case), they will fail to come to an agreement under some circumstances (see Fudenberg and Tirole, 1992). Such a failure could threaten the stability of the power system.

These transactions costs can be reduced if the Agent contracts forward: extending the contracting horizon increases flexibility and thereby mitigates holdup and bargaining problems. By extending the time available to find alternative trading partners, the ability of any one agent to hold up others is reduced and bargaining is more competitive. Any individual actor's power to extort from others is reduced to the extent that there are more substitutes available for the service that actor supplies.

For instance, the Agent can contract forward for "ancillary services" such as generating reserves.[8] Ancillary service contracts essentially involve an exchange of control rights. Owners of generating capacity transfer some of the rights to control their assets to the ancillary services purchaser. For instance, if the Agent purchases so-called spinning reserves from the owner of a thermal generating facility, the generation owner is obligated to keep the turbine spinning and synchronized with the grid; in this condition it is available to produce energy on short notice. Moreover, the Agent has the right to require the generator to produce power and provide it to the grid. By purchasing such reserves in advance, the Agent can respond to an event, such as a forced outage at another generator, by exercising the control rights transferred under the ancillary services contract.[9]

Contracting for ancillary services on a relatively short-term basis (e.g., for the next day) offers advantages because this permits the contracting parties to contract contingent upon more information with relatively simple contracts. For instance, day-ahead forecasts of power demand, weather, and resource availability are more accurate than week-ahead, month-ahead, or year-ahead forecasts. If all contingencies could be foreseen and costlessly specified, complete, long-term ancillary services contracts would be feasible. However, given the complexity of power systems, identifying actions under all relevant circumstances that may occur over long time horizons is extraordinarily costly. Sequential short-term contracting reduces transactions costs.

The immense complexity of a power system implies that forward contracting cannot eliminate all temporal specificities. Forward contracts that transfer to the Agent control rights adequate to respond efficiently to most events will not permit the Agent to respond efficiently to all events, and Murphy's Law tells us that such unprepared-for events will occur. When they do, temporal specificity and the associated quasi-rents and transactions costs will rear their ugly heads.

This problem can be addressed by providing the Agent with broad residual control rights. For example, the Agent may have the authority to curtail loads or adjust generation in the event of a contingency that threatens reliability without the permission of affected customers or the owners of relevant assets. Due to the complexity of the power system, the specific contingencies that constitute an emergency are impossible to delineate *ex ante*. Delegation of such broad and imprecisely defined authority to be employed "in the breach" has benefits and costs. On the benefit side of the ledger, the delegation of such control rights permits the Agent to prevent holdups by the owners of resources needed to respond efficiently to disruptions of the system. On the cost side, such rights may give the Agent the power to hold up the owners of generation assets. This problem may be mitigated by forcing the Agent to bear the costs associated with the exercise of these rights. Alternatively, the Agent's exercise of these rights may be subject to review by third parties (e.g., the courts). Moreover, as discussed below, the Agent's organizational form can attenuate his incentives to act opportunistically.[10]

It should be noted that the magnitude of transactions costs, and hence the importance of the coordinating Agent, are technology dependent. Transactions costs will be high, and a coordinating Agent is particularly important when (a) most electricity is generated by large-capacity central power stations and consumed by spatially dispersed loads, and (b) the capacity of the transmission system is frequently constrained. Technological innovations, such as distributed generation (where power is produced by small units co-located with consumers) reduce the need for coordination of the use of the transmission system. Similarly, investment in redundant transmission capacity reduces the likelihood of transmission constraints and the associated potential for reliability problems.

Although distributed generation has received considerable attention in recent years (see chapter 9), and there have been proposals to enhance transmission capacity, it is unlikely that either technical change or increased transmission capacity will exert a large influence on the electricity transmission system in the near to medium term. Indeed, NIMBY and budgetary problems have impeded the development of new transmission capacity even in places (such as Path 15 in California) where it is needed at the same time that secular growth in electricity consumption is placing further strains on transmission systems. Thus, although there may come a day when technology and investment sharply reduce transactions costs and the benefits of central coordination, that day is not in sight. Indeed, if anything, efficient central coordination will increase in importance in coming years.

To summarize, the public-good nature of reliability in an electrical power network provides a rationale for the creation of a coordinating Agent acting on behalf of the consumers connected to the network. This Agent can contract for resources required to maintain system reliability. Moreover, it may be efficient to endow the Agent with some residual rights of control to permit him to avoid holdups when responding to events that threaten to disrupt the system.

So far, this argument provides a justification for the creation of a coordinating Agent that is (a) empowered to contract for some resources required to coordinate system activities, and (b) endowed with some residual control rights over generation, transmission, and load. When designing a power market, however, it must be recognized that the Agent is not a deus ex machina that will automatically exercise his control rights and ability to contract on behalf of others in an efficient manner. Institutional design matters because it influences the incentives the Agent faces, and hence how the Agent will behave. The next section considers how the Agent is likely to act under some alternative institutional arrangements.

A Comparative Analysis of Institutional Arrangements

A variety of institutional arrangements have been implemented or proposed that create an agent that serves the coordinating function described above. The most important are vertically integrated utilities, independent system operators (ISOs), and independent transmission companies (transcos).

Vertically integrated utilities (either investor or government-owned) with monopoly service territories were the standard institutional arrangement until recently. Vertically integrated utilities own and operate generating assets, transmission lines, and distribution facilities. These utilities typically have an obligation to provide electrical service in a franchise service territory. Moreover, in exchange for the obligation to serve, these utilities typically receive a monopoly franchise in that territory.

Vertical integration and monopoly service territories clearly address some of the problems of transactions costs discussed earlier. In particular, a utility with a monopoly on generation and transmission in a particular territory does not face holdup problems when responding to disturbances that threaten system reliability because there is no need to negotiate with other firms to coordinate the responses required to maintain system reliability. This benefit comes at a cost, however. The granting of a monopoly over coordination and generation functions in a particular geographic region creates the potential for the monopolist to exercise market power. Vertical integration of generation and transmission can exacerbate this problem because a vertically integrated firm may withhold transmission resources from competing suppliers of generation in order to increase the firm's profits from generation. Thus, the traditional means of organizing the production and sale of electricity and the provision of reliability requires more intensive and costly regulation (where costs arise from information asymmetries—see Laffont and Tirole 1993) to address market power issues.

ISOs and transcos both involve the separation of coordination and energy production functions in an attempt to facilitate competition in generation and thereby reduce regulatory inefficiencies. ISOs own neither generating assets nor transmission lines. Instead, the owners of transmission lines delegate the right to control the use of these lines to the ISO. ISOs also coordinate the operation of the power grid through the operation of balancing and ancillary service markets. Owners of generation also delegate some residual control rights to the ISO to facilitate its maintenance of system reliability. Unlike ISOs, transcos own transmission systems; like ISOs, they own no generation assets. They effectively serve the same function as ISOs, but differ in important ways. Most important, whereas ISOs are typically nonprofit corporations in which "stakeholders"—such as owners of generation and transmission assets, and consumer representatives—have voting and control rights, transcos are independent, for-profit, investor-owned and -controlled corporations subject to regulation.

What are the costs and benefits of these alternative means of coordinating the operation of a power network? Two crucial considerations affect the efficiency of these alternatives. In particular, to operate efficiently, the coordinating Agent must have the appropriate information and face the right incentives.

Two types of information are important. First, to make appropriate decisions that affect reliability, the coordinating Agent must have good information about consumers' willingness to pay for it. Second, the Agent must have good information about the costs of alternative means of responding to system contingencies.

The first informational demand is far more onerous. This is not surprising; the problem of determining the willingness to pay for public goods has vexed economists and policy makers for a long time indeed. Willingness to pay should be related to (a) the costs associated with system problems (e.g., blackouts, brownouts) and (b) consumer risk preferences. The costs are at least conceptually amenable to calculation (perhaps through simulation of how likely power system failures are and the economic costs associated with these failures). Risk preferences are far more difficult to quantify.

For the most part it appears that the operating assumption has been that the costs of reliability shortfalls are immense, and that they should be avoided at nearly any price (see chapter 10). In the first instance this has led to the creation of very demanding system operating standards under the aegis of bodies such as the National Electricity Reliability Counsel or the Western System Supply Counsel that set operating standards for coordinating Agents (be they integrated utilities, ISOs, or transcos). These bodies promulgate "prudent" operating reserve levels that essentially serve as quantity-setting "command and control" mechanisms. Implicitly these standards signal that the marginal cost of a violation of specified reserve levels is effectively infinite. Rigid adherence to these standards can lead to unsatisfactory outcomes, such as extremely high prices for ancillary services without a concomitantly valuable improvement in system reliability.

In response to such problems, ancillary service prices have been capped in some jurisdictions; caps imply that the willingness to pay for reliability is finite after all. Caps are a relatively crude mechanism that in essence imply that the demand for reliability is perfectly inelastic at prices below the cap level, but perfectly elastic at the cap price. Some Agents, such as the New England ISO, have recognized that reliability is like any other economic good and is therefore subject to downward sloping demand. These Agents are investigating the possibility of creating a mechanism that would take this reality into account (see chapter 10, Chao and Wilson 1999). This is a step in the right direction, but much work remains to be done to estimate just what the demand curve looks like. Historically, reliability decisions have been engineering-driven. Economists and economic reasoning could provide valuable input to this process. The efficiency of the power system would be greatly enhanced if decisions were predicated on better information on the value of reliability.

The second information problem is more tractable. The costs of particular responses to system contingencies depend on the costs of generators hooked up to the system, the constraints that inhibit the ability of generators to adjust output, and the configuration of the transmission network. Vertically integrated utilities that own and operate generators have precise information on costs and constraints. Bidding mechanisms can generate this information for system operators (ISOs or transcos) that do not own generation. For instance, the Agents can operate markets in which generators submit bids indicating the prices at which they are willing to increment or decrement generation. To the extent that these markets are competitive, the bids will reflect accurately the costs of such adjustments. Competitive ancillary services markets provide the Agent with information about the costs generators incur to respond to contingencies.

Whatever the information possessed by the Agent, institutional design in the power market should strive to ensure that the Agent responds to this information in an efficient way. There are two crucial elements to this design: the organizational form of the Agent and the legal environment in which it operates.

Organizational form can influence the incentives of the Agent in a variety of ways. Perhaps most important, organizational form affects the power of the incentives the Agent faces. Indeed, incentive power is a crucial issue in power markets.

Incentive power relates to the strength of the connection between an economic agent's performance and his compensation. Under a "high-powered" incentive regime there is a close connection between pay and performance; piece-rate and sales-commission systems are examples of high-powered incentives. In contrast, in a "low-powered" incentive system there is only a weak connection between performance and compensation; a fixed-wage labor contract is an example of a low-powered mechanism.

Although high-powered incentives are desirable in neoclassical settings in which transactions costs are unimportant, a wide variety of analytical approaches show that information problems can favor attenuation of incentive power.[11] These problems often arise when (a) the quality of output—such as the reliability of a power system—can be affected by the firm producing it, and (b) information about quality is not symmetric or quality cannot be contracted on. If the firm has better information about output quality than its customers (in the extreme, if the firm knows quality perfectly and consumers cannot verify quality at all), when facing high-powered incentives the firm may produce too low quality in equilibrium; this is the well-known lemons problem. Difficulties may also arise even if consumers can *observe* output quality, but cannot enter into contracts that specify quality. This may occur if quality is not verifiable by a third party (such as a court). When quality is non-verifiable, a firm facing high-powered incentives can chisel on contracted quality and thereby reduce costs without fear of legal consequences. In certain settings (e.g., single-shot contracting games) consumers have no recourse against such chiseling, and the only equilibrium involves production of low-quality goods even though consumers would be willing to pay for high-quality ones (Easley and O'Hara 1983; Glaeser and Shleifer 1998).

These problems can be mitigated by reducing incentive power, that is, by reducing the strength of the connection between compensation and performance. In particular, theoretical research shows that nonprofit firms may produce more efficiently than for-profits when producers have superior information about quality (Easley and O'Hara) or when quality is noncontractible (Glaeser and Shleifer). Reducing incentive power reduces the benefits the firm can capture by chiseling on quality; it cannot pocket the associated cost savings. Thus, nonprofit firms may produce higher quality output when quality is (or important attributes thereof are) noncontractible or nonobservable by consumers. Although reducing incentive power through the formation of nonprofit firms leads to inefficiencies on other dimensions (such firms may produce observable or contractible quality less efficiently than their for-profit counterparts because the former cannot capture the benefits of the higher quality by charging a higher price and pocketing the resulting profit), if the nonobservable or noncontractible elements of quality are sufficiently important nonprofit firms may be more efficient than for-profit ones.

This analysis is directly relevant for the power industry because the complexity of the power grid plausibly creates nonobservability or noncontractibility problems. As long as the lights go on, consumers may be blissfully unaware if the coordinating Agent is chiseling on reliability. Moreover, chiseling on reliability may have no consequences under most circumstances but may have disastrous effects with small probability. The coordinating Agent almost certainly has far better information about this probability than power consumers. Noncontractibility is also almost certainly present. Contracts will be incomplete due to the complexity of grid operation. If the Agent promises to take the actions necessary to ensure optimal reliability, these promises are almost certainly unenforceable in court. If consumers sue the Agent, claiming that he has chiseled on the promised level of quality, courts will have a very difficult time ascertaining whether this is in fact true because they will not have the requisite knowledge about the operation of a complex power system to make an informed judgment.

Under these circumstances, a strong case can be made for subjecting the Agent to a low-power incentive regime. For instance, requiring the Agent to be a nonprofit enterprise can provide an incentive for him to deliver a more efficient level of reliability than a for-profit. ISOs are nonprofit. Moreover, transcos operating subject to rate-of-return or profit regulation also face attenuated incentive power.[12] These arrangements are sensible given the complexity of power systems and the associated implications for observability and contractibility. Low incentive power can also reduce the incentive of the coordinating Agent to hold up users of the transmission system because he cannot capture the associated gains. This reduces the cost of transferring residual control rights.

There are other mechanisms for addressing quality problems. For instance, even though consumers may not be able to observe quality in real time, they may learn it with a lag; reliability may be an experience good. If they deal repeatedly with the Agent, they can punish an Agent that chisels on quality by paying lower prices in the future even if quality is noncontractible. Under these circumstances, repeat dealing and reputation can deter chiseling. Some economists (e.g., Zycher 2000) argue that a reputational mechanism can motivate for-profit Agents to provide (nearly) optimal reliability.

There is considerable room for skepticism regarding these claims. In particular, power system reliability is substantially different from other experience goods. The regular consumer of soft drinks who finds that too many cans sold by a particular firm are flat because of inefficient quality control efforts pays a relatively low price for this information; the same could not be said of a storeowner whose shop is looted during a blackout. That is, reputation is less efficient as a disciplining method when the consumer only learns about the chiseling with small probability and incurs an extreme cost when he does so. This is plausibly the case in the power industry.[13]

Put differently, reputation is costly. To generate a valuable reputation, firms may have to price in excess of marginal cost (thereby distorting consumption and production) to generate a stream of cash flows that would be forgone in the event consumers discover chiseling on quality; when the probability of discovery is low and the cost incurred upon discovery is high, this price-marginal cost wedge may have to be extremely large to provide the appropriate incentive. In this case, it could be better to implement a low-power incentive regime rather than rely on high-powered reputational incentives; this is a plausible outcome in power markets. Alternatively, firms may invest in specific assets or hold large quantities of equity as "hostages" to incent them to supply high-quality output. These are also costly. Again, these reputational costs may be higher than the costs associated with low-powered incentives in circumstances that are plausible in power markets.

The complexity of the power system and the potentially huge costs associated with a system failure likely make reputation a very expensive mechanism for providing an incentive to provide reliability. Due to the low probability of failure, the high cost thereof, and the difficulty of monitoring the actions of the reliability supplier, the power system is not a good example of an experience good for which reputation is a cheap and efficacious means of ensuring quality.

Legal rules can also play an important role in motivating the Agent to supply reliability efficiently. Making the Agent liable for system failures could provide valuable incentives. In particular, strict liability would force the Agent to internalize the costs of the decisions that affect reliability. This would have the collateral benefit of giving the agent an incentive to estimate these costs precisely, which would lead to better reliability decisions (see the earlier discussion of information issues). Strict liability has advantages in this context relative to alternatives such as negligence. A negligence rule is likely to founder on the same complexities that plausibly make quality noncontractible in power markets. Specifically, the complexity of power system operations makes it very costly for third parties to determine whether the coordinating Agent acted negligently.

Strict liability requires no such extensive fact-finding; if a failure occurs, the Agent is liable regardless of the actions he took to avoid it. The main difficulty with a strict liability regime in this case is judgment-proofness. Given the large costs that could result from suboptimal supply of reliability, an Agent would chisel if its equity is smaller than these costs because it knows that it would bear only a fraction of the costs associated with an actionable system failure. Therefore, relying on liability to motivate the Agent would require the Agent to hold substantial equity. Since equity capital is costly (due to information asymmetries), a liability regime also entails costs. Due to equity costs and judgment-proofness, it may be necessary to attenuate the power of incentives facing the coordinating Agent and to subject him to some *ex ante* regulation of his activities.

The foregoing suggests that the characteristics of power systems pose acute challenges to any institutional framework. Traditional quality-assurance mechanisms such as reputation and liability are likely to be very costly in power markets. Under these circumstances, it is likely necessary to attenuate incentive power in order to provide reliability efficiently. Thus, nonprofit ISOs and profit-regulated transcos are sensible institutional arrangements in the power market.

This said, although both ISOs and profit-regulated transcos face low-powered incentives, they are not necessarily perfect substitutes for one another.

Indeed, there is considerable controversy about which mechanism is superior. This debate has too many facets to address here, so I will restrict my attention to one: governance.[14] Nonprofit ISOs have no residual claimants (who usually exercise governance/control rights in for-profit firms), so most ISOs are governed by stakeholders who have voting rights but no residual claim on cash flows. These stakeholders include owners of generating capacity in the ISO's territory, owners of transmission capacity, firms that have obligations to service load, power marketers, and consumer representatives.

It is well recognized that these stakeholders may have an incentive to distort ISO decisions for their own benefit (Michaels 2000). For instance, generation owners may try to dissuade the ISO from undertaking actions that improve reliability but erode their market power; as an example, a generator may oppose an effort to build a new transmission line that increases the number of other generators that can sell power in his geographical region. Michaels also notes that ISOs may face problems because, as first noted by Arrow in his "impossibility theorem," collective organizations such as ISOs lack a well-behaved (e.g., transitive) objective function. This can make ISO decision making unstable and unpredictable. Moreover, ISOs are political institutions, and stakeholders will seek rents through their influence on ISO decision making.

Michaels argues that an investor-owned transco faces no such problem. A profit maximizing transco has one objective—to maximize profits—that eliminates the collective action problems inherent in the nature of an ISO.

Michaels is correct that for-profit transcos have a simple objective—to maximize profits for their owners. The analysis presented in this chapter implies, however, that this is likely a liability rather than an asset in the power industry; due to this simple objective function, transco operations may be predictable, but this is no guarantee that they are predictably superior. High-powered incentives can compromise reliability. Just because the for-profit corporate form has proved optimal in other circumstances (a point emphasized by Michaels) does not imply that it is the optimal arrangement for supplying reliability in power markets. As Hansmann (1996) shows, a wide variety of "non-standard" organizations—that is, organizations other than the for-profit corporation—have evolved and survive because they address economic problems specific to a particular industry or good.

A more detailed analysis of the organizational challenges of supplying reliability in power markets suggests that a nonprofit ISO structure has desirable features as compared to for-profit alternatives. Although rent seeking through manipulation of the governance process is clearly a cost of the ISO form, there are important benefits too. For instance, giving stakeholders voting rights within the ISO can mitigate other forms of rent seeking, such as holdups.[15]

One group of stakeholders can use their voting rights to block wasteful holdup attempts. Moreover, it may be cheaper to negotiate, exercise, and enforce the broad residual control rights required to protect reliability in the breach (see above) in the context of a cooperative structure where the affected parties have voting rights. In contrast, independent investor-owned transcos may attempt to hold up generation owners and at the same time be vulnerable to holdups by owners of generation. Negotiation, exercise, and enforcement of the in the breach control rights may also be more difficult across different firms than within a formal organization in which affected parties have a vote. Furthermore, like other political organizations, nonprofit firms can—and do—devise procedural rules to mitigate preference instability (Pirrong 2000).[16] A priori, it is not clear which organizational form dominates because they are both vulnerable to rent seeking, albeit on different dimensions.

Experience from another industry suggests that the nonprofit form may have some important benefits. In particular, traditional financial exchanges share many features with ISOs. Both are nonprofits with heterogeneous members with disparate interests. Both have been criticized for their politicized governance (as noted earlier, Michaels criticizes the political nature of nonprofit ISOs). Both have been accused of providing a venue for wasteful rent seeking. Exchanges are clearly guilty as charged. Pirrong (1995a, 1995b, 2000) shows that exchanges have often adopted inefficient rules (such as commission cartels and weak rules against manipulation) because they benefited a politically powerful group of exchange members.

Despite all of their failings, however, nonprofit exchanges have proved an extremely durable form of economic organization. They have passed the survivor test. They have done so because, to paraphrase Churchill, they are the most inefficient form of organization ever designed to govern trading of financial assets, with the exception of every other that has been tried. They have flourished despite all of their recognized failings because they have proven the most efficient means of mitigating (but not eliminating) the conflicts that arise from the heterogeneity of the agents involved in trading financial assets. This heterogeneity inheres in the technology of trading; given the nature of human capital used in traditional floor trading, integration is a very costly mechanism for aligning incentives, and consequently numerous heterogeneous and independent agents provide trading services on an exchange. Political governance and nonprofit form are an efficient adaptation to this heterogeneity. The nonprofit form also attenuates the incentive of stakeholders to manipulate the exchange's pricing and payout policies to seek rents (Pirrong 2000).

The experience of exchanges suggests that inventorying the failings of ISOs is not sufficient to demonstrate their inferiority as a means of coordinating the operation of power markets. When the fundamental heterogeneity of participants in a dis-integrated power market is recognized, the history of exchanges suggests that a nonprofit, cooperative form of organization such as an ISO may be more efficient than alternatives, such as investor-owned, for-profit transcos. Given the complexity of pricing of transmission services and the joint-production inherent in the generation and transmission of power, rent seeking through manipulation of pricing and payout policies is likely to be a serious concern; the nonprofit form addresses this concern to some degree. The cooperative form and low-powered incentives inherent in the nonprofit ISO model also plausibly reduce the costs associated with other conflicts inherent in a dis-integrated power industry.

It should also be noted that the ISO is not a new concept devised by the fevered imagination of some academic scribbler or politician. Power pools such as NEPOOL and PJM—the forerunners of ISOs—developed spontaneously from the efforts of utilities to address reliability issues; utilities recognized the benefits of cooperating to address reliability and coordination concerns. Although, as critics have noted—and as I have acknowledged above—these organizations have often made self-serving decisions, this does not imply that the ISO concept is inherently inferior to alternatives. Given the nature of the electricity system and the great diversity of the participants who affect and are affected by the reliability of its operation, an organization facing low-powered incentives that gives voting and control rights and voice to these disparate parties has much to recommend it.

SUMMARY AND CONCLUSIONS

The power industry is undergoing profound changes around the world. These changes raise fundamental questions about the efficient extent of decentralization in power markets, and the appropriate organizational and legal framework in which this decentralization should take place.

Transactions-cost and property-rights economics have important things to say about these issues. Fundamental transactions-cost considerations imply that extreme decentralization is not feasible in power markets. Due to the public good nature of system reliability and the short operational time frame in which adjustments to shocks must occur, there is a role for a central coordinating Agent responsible for operating the transmission grid to maintain reliable operation of the power system. Moreover, mitigation of holdup problems requires endowing the Agent with some residual control rights over generating and transmission assets.

Creation of a coordinating Agent is not sufficient to ensure efficiency. This Agent must possess the relevant information and face the right incentives to act efficiently. At present, there is little reason to be confident that coordinating Agents have the information necessary to determine the efficient level of reliability. This is a promising area for future research.

Incentive issues have received more attention. In particular, the organizational form of the coordinating Agent, and the consequent incentive effects, are matters of intense debate. Fundamental transactions-cost considerations suggest that subjecting the Agent to low-powered incentives (such as those inherent in a nonprofit form of organization or profit regulation) is desirable in order to motivate him to supply reliability efficiently. Moreover, the heterogeneity of participants in a dis-integrated, but inherently interconnected, power industry suggests that nonprofit cooperative organizations that extend voting rights to participants at all levels of the production chain have important advantages over alternative forms of organization such as for-profit transcos.

Recognition of the complexity of the power industry dissuades me from stating flatly that nonprofit ISOs are superior, but my research on financial exchanges shows that cooperative, nonprofit organizations can be an efficient means of governing transactions in network industries with heterogeneous agents. The dis-integration of the electricity industry that is the centerpiece of restructuring around the world has created an environment in which heterogeneous agents supply a joint product. This situation is analogous to that facing traditional financial exchanges. Exchanges employ nonprofit cooperative organizational forms and political governance to mitigate the inherent conflicts. This approach is arguably the best way to organize the coordinating function in power markets, although it must be admitted that the argument is not clinched. More work needs to be done. Moreover, given our limited understanding of the relevant trade-offs, experimentation in organizational form should be encouraged.

REFERENCES

Centolella, P. 1996. The Organization of Competitive Wholesale Markets and Spot Price Pools. The National Council on Competition and the Electric Industry.

Chao, H., and H. Huntington. 1998. *Designing Competitive Electricity Markets*. Cambridge, MA: Kluwer Academic Publishers.

Chao, H., and R. Wilson. 1999. Design of Wholesale Electricity Markets. Electric Power Research Institute working paper.

Coase, R. 1937. The Nature of the Firm. *Economica* 4: 386–405.

———. 1960. The Problem of Social Cost. *Journal of Law and Economics* 3: 1–44.

Crocker, K., and S. Masten. 1996. Regulation and Administered Contracts Revisited: Lessons from Transaction-Cost Economics for Public Utility Regulation. *Journal of Regulatory Economics* 9: 5–39.

Easley, D., and M. O'Hara. 1983. The Economic Role of the Non-Profit Firm. *Bell Journal of Economics* 2: 531–38.

Ellerman, A. D., P. L. Joskow, R. Schmalensee, J.-P. Montero, and E. M. Bailey. (2000) *Markets for clean air: The U.S. acid rain program*. Cambridge: Cambridge University Press.

Fudenberg, D., and J. Tirole. 1992. *Game Theory*. Cambridge, MA: MIT Press.

Glaeser, E., and A. Shleifer. 1998. Not-for-Profit Entrepreneurs. NBER working paper W6810.

Hansmann, H. 1996. *The Ownership of Enterprise*. Cambridge, MA: Harvard University Press.

Holmstrom, B., and P. Milgrom. 1991. Multitask Principal-Agent Analyses: Incentive Contracts, Asset Ownership, and Job Design. *Journal of Law, Economics, and Organization* 7: 24–42.

Joskow, P. L. and E. Kahn. (2001) *A Quantitative Analysis of Pricing Behavior In California's Wholesale Electricity Market During Summer 2000*. National Bureau of Economic Research Working Paper 8157.

Joskow, P., and R. Schmalensee. 1983. *Markets for Power: an Analysis of Electric Utility Deregulation*. Cambridge, MA: MIT Press.

Klein, B. and K. Leffer. 1981. The Role of Market Forces in Assuring Contractual Performance. *Journal of Political Economy* LXXXIX(4): 615–41.

Laffont, J-J., and J. Tirole. 1993. *A Theory of Incentives in Procurement and Regulation*. Cambridge, MA: MIT Press.

Masten, S., J. Meehan, and E. Snyder. 1991. The Costs of Organization. *Journal of Law, Economics, and Organization* 7: 1–26.

Michaels, R. 2000. Can Nonprofit Transmission Be Independent? *Regulation Magazine* 23: 61–66.

Pirrong, C. 1993. Contracting Practices in Bulk Shipping Markets: A Transactions Cost Explanation. *Journal of Law and Economics* 36: 937–75.

———. 1995a. The Efficient Scope of Private Transactions-Cost-Reducing Institutions: The Successes and Failures of Commodity Exchanges. *Journal of Legal Studies* 24: 229–55.

———. 1995b. The Self-Regulation of Commodity Exchanges: The Case of Market Manipulation. *Journal of Law and Economics* 38: 141–206.

———. 2000. A Theory of Financial Market Organization. *Journal of Law and Economics* 43: 437–72.

Weingast, B. R. and W. J. Marshall. 1988. The Industrial Organization of Congress or Why Legislators, like Firms, Are Not Organized as Markets. *Journal of Political Economy*. 96(1): 132–63.

Williamson, O. E. 1985. *The Economic Institutions of Capitalism*. New York: Free Press.

Zycher, B. 2000. Keeping the Power On. *Regulation Magazine* 4: 7–11.

NOTES

1. See Centolella (1996) for a succinct description of the operation of an electric power system.

2. Power plants have a maximum amount of power that they can generate (assuming that they have sufficient fuel input). A 500 megawatt plant that is not energy input (fuel) constrained can generate 500 megawatts of electricity. As an example of an energy input constraint, the output of a hydroelectric plant may be limited by the water available to run over the dam.

3. In the United States clean air regulations restrict the output of sulfur dioxide (SO2). There is also a program for power plants to trade SO2 permits granting the right to emit a given quantity of the pollutant Ellerman et al. (2000). In California, the state restricts output of nitrous oxide (NOx) and has implemented a similar permit program for this pollutant Joskow and Kahn (2001).

4. See Masten, Meehan, and Snyder (1991), and Pirrong (1993) for a discussion of temporal specificities and transactions costs.

5. A quasi-rent is the difference between the value of a resource in its first-best use and the payment required to keep it in operation.

6. Crocker and Masten (1996) also note that temporal specificities are likely to be important in power markets, and that these specificities can create substantial transactions costs.

7. See especially Williamson (1985).

8. Reserves are generating resources precommitted to respond to the coordinating Agent's instructions within a specified time frame. This generating capacity can be used to respond to demand shocks, forced outages, or transmission line failures. Suppliers of reserves incur an opportunity cost; resources supplied as reserves cannot be used to generate energy.

9. Due to the spatial distribution of generation and loads, there may be poor substitutes for some generation units as suppliers of reserves. Market power problems may exist for these "reliability must run" units. Indeed, because power networks were created by vertically integrated utilities that did not need to consider reliability-driven holdup problems, many generating units may have poor substitutes for reliability-maintenance purposes under common grid conditions. Thus, holdup problems are likely ubiquitous in formerly vertically integrated power systems.

10. Different market designs have conferred different control rights on coordinating Agents. For instance, the California ISO (the coordinating Agent in that state) has sharply circumscribed residual control rights. In contrast, the Agents in "tight" power pools such as NEPOOL and PJM have extensive residual control rights. The early experience in California suggests that California ISO has been subject to some holdup problems. Similar problems have not been experienced in tight pools, but the Agents in these pools have been criticized for heavy-handed intervention, which could be construed as holding up generators. Thus, the allocation of residual control rights to coordinating Agents appears in fact to influence holdup-related transactions costs.

11. See especially Williamson (1985) and Holmstrom and Milgrom (1991) for a discussion of the costs and benefits of high-powered and low-powered incentive regimes.

12. Other forms of transco regulation may be counterproductive. In particular, price caps provide a high-powered incentive and could be extremely counterproductive under the conditions prevailing in the power market.

13. A small probability of detection of chiseling effectively increases the relevant discount factor. Laffont and Tirole (1993) present a model in which a high discount factor can make reputation an inefficient means of providing incentives to produce quality. See also Klein and Leffler (1981).

14. One largely unresolved issue relates to investment in new transmission capacity. Organizational form and property rights almost certainly affect incentives to invest in capacity. This chapter focuses on short-run issues rather than such long-run ones. Pricing of transmission—most notably, pricing of congested transmission resources—is also a matter of considerable controversy. These issues are discussed in chapter 7 and several chapters of Chao and Huntington (1998).

15. The effect of voting rights on holdups depends in part on the structure of these rights. Obviously, a unanimity rule makes holdups more likely. However, a majority rule or some supermajority rule in which no single participant's vote is likely to prove pivotal can mitigate holdup problems.

16. See also Weingast and Marshall (1988) for an analysis of how rules in the U.S. Congress serve to mitigate preference instability, voting cycles, and other collective-action problems.

7

Market-based Transmission Investments and Competitive Electricity Markets[1]

William W. Hogan

$\mathbf{M}$arket-based transmission investments can play an important role in competitive electricity markets. A short-term electricity market coordinated by a system operator provides a foundation for a competitive electricity market. In this setting, locational price differences define the opportunity cost of transmission. The potential to arbitrage these same price differences provides a market incentive for transmission investment if there is a method to capture the benefits of investment. For an integrated grid, transmission congestion contracts are equivalent to perfectly tradable physical transmission rights. With such contracts to allocate transmission benefits, it would be possible to rely more on market forces, partly if not completely, to drive transmission expansion.

INTRODUCTION

A framework for market-based investments in transmission would extend the scope of competitive electricity markets. Electricity market restructuring emphasizes the potential for competition in generation and retail services, with the operation of transmission and distribution wires as a monopoly. Network interactions and economies of scale both complicate the extension of market incentives to investment in the wires business. Two broad approaches suggest themselves for dealing with these network problems in a manner compatible with a competitive generation market.

There could be monopoly management of transmission operations and investment, with incentive pricing for the monopoly. Transmission would be like a large "black box" run by the monopoly, which takes on an obligation to provide unlimited transmission service for everyone. With the appropriate price cap or other incentive regulation, the monopoly would make efficient investments or contract with market participants to remove or manage the real transmission limitations. This approach results in a powerful monopoly with the familiar problems of finding the right level and form of incentive regulation. It is found, for example, in the market in England and Wales with the National Grid Company.[2] Although some elements of this monopoly approach may be found in any practical framework for transmission investment, it is not the focus here.

An alternative approach would lean more in the direction of market mechanisms. This would allow market participants to make transmission investments in response to price incentives. The transmission investment would be voluntary, rather than included in mandatory charges. The benefits would be captured through tradable transmission property rights.[3] A workable system that provides an equivalent to transmission property rights could be used by market participants to guarantee the costs of the actual flow of power or traded in a secondary market. Acquisition of these transmission benefits would provide a market incentive for transmission investment.

Market-based transmission investments confront a number of challenges. In practice, the difficulties can lead to a conundrum, a Catch-22. The typical example of a transmission investment invokes the image of a large new transmission line, which might look to be too difficult to base solely on market decisions. As a result, the implicit assumption often is that only a regulated monopoly could manage the intended investment, and the discussion of market institutions defaults to the design of monopoly mechanisms. Market-based investments may never get a hearing and would be foreclosed in the market design. However, with some innovation it is possible to envision at least a partial escape from this trap by designing market institutions that could meet some or all of the challenges and support market-based transmission investments. The purpose here is to outline such a framework.

A central problem is in defining a workable notion of property rights for transmission. In the case of simple physical rights and controllable transmission flows, market-based investments would be easy to define and a natural approach for transmission expansion. In this case, analogies to other markets would apply. In the real electricity market, however, physical rights for the full capacity of the grid are difficult to define, and controllable flows are the exception rather than the rule. However, it is still possible to define a workable system of financial transmission rights that could achieve the same outcomes and could support a system of market-based investments, using the foundation of an electricity market run by a system operator.

CONTROLLABLE FLOWS AND TRANSMISSION RIGHTS

If the world were simple, market-based investments in transmission would provide a natural and self-evident approach to transmission expansion. Suppose that an electric transmission network consisted only of transmission lines and valves. In this easy case, the power would flow down the lines from source to sink, and the valves could make the system completely and continuously controllable in the sense that the actual path of power flows could be assured, no matter what the pattern of power inputs and outputs in the network. In principle, power flows could be labeled, directed, and tracked. We could charge directly for the power flows on each line. If we did not want the power to flow down a particular line, the valve could be closed for those who did not pay. In this world, there would be no network externalities. The convenient contract-path model of electric power transmission would apply. The owner(s) of a line could charge for its use. In competitive equilibrium, the price of that usage would be equal to the difference in the prices of electricity at the source and sink. Equivalently, the owners of the line could buy at the source and sell at the sink, profiting from the difference in locational prices.

Suppose further that investments in transmission lines came in arbitrarily small increments with constant or increasing marginal costs. In other words, investment was not lumpy, and there were no economies of scale. Then the transmission line would be like a bundle of straws, and we would be able to expand the transmission line by adding the cheapest available straw, with anyone allowed to make the investment. The investment would be profitable as long as the difference in locational prices between source and sink exceeded the cost of the incremental straw. In equilibrium, the cost of the last straw added would just equal the resulting expected difference in prices. The profits collected from the purchase and sale of power would be just enough to pay for the market-based investment. Furthermore, in equilibrium all the early investments in cheaper straws would enjoy a capital gain equal to the difference in this market clearing price and the cost of constructing their particular straw.

Under these conditions in a competitive market, when equilibrium prices differed at the source and sink, there would be three important equations. First, the amount of power purchased and sold at source and sink would equal the capacity created by the transmission investment. Second, the power flowing on the controllable transmission line would be equal to this same capacity. And third, the expected price difference between the locations would be equal to the marginal cost of expansion.

In this simple world with no network externalities or economies of scale, market-based investments alone would be able to achieve the efficient outcome. Investors would follow the price signals and their own profit incentives. The equilibrium difference in locational prices would just cover the cost of the marginal investment. The resulting level of investment in transmission would be just right, given the total costs of investment and the value of additional capacity.[4] The market would work efficiently.

Unfortunately, this simple world has little in common with the more complex reality of the electric transmission system. Just a few of the exceptions have an important impact on the models that can work to support market-based transmission investment. The very nature of interconnected free-flowing transmission networks creates powerful network externalities. Most lines are not controllable, and power flows do not follow the designated contract path. Rather we have the familiar reality that power flows across every parallel path between source and sink. In addition, the services provided by the transmission system go well beyond the model of a straw between two points. For instance, reactive power availability can have a significant impact on the transmission capacity to move real power between many locations. This means that investments such as, for example, capacitor additions to provide reactive support can increase the capacity of the transmission system without constructing new lines on which the power would flow. Furthermore, it is a commonplace that there can be significant economies of scale in transmission expansion. Hence, the technically efficient level of investment might come in lumps that change equilibrium prices substantially.

Thus, a fully efficient transmission investment could destroy at least two of the important equations above. The amount of power purchased and sold at the source and sink could still equal the expanded transmission capacity, properly defined. However, the power actually flowing over the "line" need not be the same. Furthermore, with substantial economies of scale, we might eliminate the important linkage between expected price differences and the marginal cost of investment; the expected price difference between the locations need not be equal to the marginal cost of expansion.

In this real world, power flows across individual lines would not be equal to the increments in capacity, and the simple model of charging for the flows on the line would not support efficient investments. In the extreme case, with economies of scale and the efficient investment eliminating the price difference between locations, there might be no investment at all.

The response to these well-known facts is generally as follows: Given the economies of scale and scope, and the large network externalities, the assumption is that only a monopoly could solve the problem of transmission investment. The monopoly, of course, would not need to rely on the voluntary investment choices of market participants. The monopoly could make the investment, under regulatory supervision, and then send the bill to the market participants. Here, however, we look beyond acceptance of the monopoly solution to examine the needed innovations and consider the role of market-based investments in transmission expansion.

A ROLE FOR MARKET-BASED TRANSMISSION INVESTMENTS

While there are circumstances where the monopoly solution may be the only practical alternative, innovations in market design and policy can address some of the most important special characteristics of the transmission grid. This requires a different approach and perspective, but the innovations would allow for some, perhaps substantial, reliance on market-based transmission investments. Here we outline the most prominent of these issues, first for the treatment of economies of scale. Then we turn to innovations to deal with the problems of network externalities, parallel flows, and economies of scope.

First, ignore the effect of network interactions and focus only on the cases where these problems are avoided and a workable system of transmission rights is in place. In the simplest case, these transmission rights would arise from controllable lines. Or the network equivalent may be assumed. Then market-based investments would cover many situations. Locational differences in prices would provide the market incentive for transmission expansion. In the case of relatively small, modular expansions of transmission capacity, the incremental transmission rights could provide the essential ingredient to complete the investment. These small investments would not have a material effect on prices. Hence, the equilibrium prices expected after the investment could be sufficient to pay for the investment. Then there would be no need to look beyond the obvious, and the market could handle this form of transmission investment, given the opportunity.

Even in circumstances where the transmission expansion might have a material impact on prices, there could be situations where the market investment would go forward. To the extent that the transmission component was part of a larger package that involved generation and contracts with customers, the cost of transmission might not dominate the investment decisions. Here the combined acquisition of transmission rights to a location and a purchase power contract at the location could support a market-based investment, given the opportunity.

In the case of larger investments, the problem of economies of scale would appear. The efficient investment might have such an impact on equilibrium prices that there would be no opportunity to pay for the investment once made. The option of contracting might not be available, because even though the investment is efficient, everyone prefers to avoid payment and free-ride on the investment of others. This is the typical case that supports the judgment that market-based transmission investment would be inadequate.

A practical compromise in this latter case would be to limit access to the new capacity (limit power flows across the new capacity), for at least long enough to justify the investment. An ability to control use of the incremental capacity could preserve sufficient margins in the locational price differences to justify the transmission investment as a stand-alone business. This would make the transmission investment like market-based investments in other industries where there are economies of scale. Furthermore, the investment could provide other benefits through relief of contingency constraints or provision of voltage support to include expansions of transmission capacity in cases other than the construction of new transmission lines. Although this would not be the "first best" efficient investment, we would be better off in the aggregate compared to the case where the investment was not forthcoming.

Those working to design market institutions should ensure that the market-based transmission investment option is included, not foreclosed by an unexamined judgment that only regulated monopoly solutions would work. In theory, market-based transmission investments have a lot to offer. In practice, they have already begun. Given the opportunity, market-based transmission investments can play an important, if not exclusive, role in transmission expansion.

ECONOMICS OF A COMPETITIVE ELECTRICITY MARKET

A general framework that encompasses the essential economics of electricity markets provides a point of reference for evaluating market design elements (see figure 7.1). Here we focus on the implications for competition in generation, and the rules for the wholesale market. The treatment of competition for other contestable elements, such as retail services, is important but need not affect the design of the wholesale market. This framework provides a background for evaluating the prescriptions for independent system operators (ISO) and related market institutions that could support market-based transmission investments.

Competitive Market Design

Reliable operation is a central requirement and constraint for any electricity system. Given the strong and complex interactions in electric networks, cur-

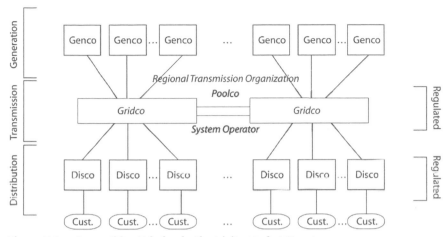

Figure 7.1. Competitive Wholesale Electricity Market Structure

rent technology with a free-flowing transmission grid dictates the need for a system operator that coordinates use of the transmission system. Control of transmission usage means control of dispatch, which is the principal or only means of adjusting the use of the network. Hence, open access to the transmission grid means open access to the dispatch as well. In the analysis of electricity markets, therefore, a key focus is the design of the interaction between transmission and dispatch, in both procedures and pricing, to support a competitive market.

To provide an overview of the operation of an efficient, competitive wholesale electricity market, it is natural to distinguish between the short-run operations coordinated by the system operator and long-run decisions that include investment and contracting. Market participants are price takers and include the generators and eligible customers. For this discussion, distributors are included as customers in the wholesale market, operating at arm's length from generators. The system is much simpler in the very short run, when it is possible to give meaningful definition to concepts such as opportunity cost. Once the short-run economics are established, the long-run requirements become more transparent. Close attention to the connection between short- and long-run decisions isolates the special features of the electricity market.

Short-run Market

The short run is a long time on the electrical scale, but short on the human scale, say, half an hour. The short-run market is relatively simple. In the short run, locational investment decisions have been made. Power plants, the transmission grid, and distribution lines are all in place. Customers and generators are connected, and the work of buyers, sellers, brokers, and other service entities is largely complete. The only decisions that remain are for delivery of power, which in the short run is truly a commodity product.

On the electrical scale, much can happen in half an hour, and the services provided by the system include many details of dynamic frequency control and emergency response to contingencies. Due to transaction costs, if nothing else, it would be inefficient to unbundle all of these services, and thus many are covered as average costs in the overhead of the system. How far unbundling and choice should go is an empirical question. For the sake of the present discussion, we focus on real power and assume that further unbundling would go beyond the point of diminishing returns in the short-run market.

Over the half hour, the market operates competitively to move real power from generators to customers. Generators have a marginal cost of generating real power from each plant, and customers have different quantities of demand depending (perhaps) on the price at that half hour. The collection of generator costs stacks up to define the generation "merit order," from least to most expensive. This merit order defines the short-run marginal-cost curve for the market, which governs power supply. Similarly, customers have demands that are sensitive to price, and higher prices produce lower demands. Power pools provide the model for achieving the most efficient dispatch given the short-run marginal costs of power supply. The system operator controls operation of the system to achieve the efficient match of supply and demand.

Given enough competitors and no collusion, the short-run economic dispatch market model can elicit bids from buyers and sellers. The system operator can treat these bids as the supply and demand and determine the balance that maximizes benefits for producers and consumers at the market equilibrium price. Hence, in the short run electricity is a commodity, freely flowing into the transmission grid from selected generators and out of the grid to the willing customers. Every half hour, customers pay and generators receive the short-run marginal cost (SRMC) price for the total quantity of energy supplied in that half hour. Everyone pays or receives the true opportunity cost in the short run. Payments follow in a simple settlement process.

Transmission Congestion

This overview of the short-run market model is by now familiar and found in operation in many countries. However, this introductory overview conceals a critical detail that would be relevant for transmission pricing. Not all power is generated and consumed at the same location. In reality, generating plants and customers are connected through a largely free-flowing grid of transmission and distribution lines.

In the short run, transmission too is relatively simple. The grid has been built and everyone is connected with no more than certain engineering requirements to meet minimum technical standards. In this short-run world, transmission reduces to nothing more than putting power into one part of the grid and taking it out at another. Power flow is determined by physical laws, but a focus on the flows—whether on a fictional contract path or on more elaborate allocation methods—is a distraction. The simpler model of input somewhere and output somewhere else captures the necessary reality. In this simple model, transmission complicates the short-run market through the introduction of losses and possible congestion costs.

Transmission of power over wires encounters resistance, and resistance creates losses. Hence, the marginal cost of delivering power to different locations differs at least by the marginal effect on losses in the system. Incorporating these losses does not require a major change in the theory or practice of competitive market implementation. Economic dispatch would take account of losses, and the market equilibrium price could be adjusted accordingly.

Transmission congestion has a related effect. Limitations in the transmission grid in the short run may constrain long-distance movement of power and thereby impose a higher marginal cost in certain locations. Power will flow over the transmission line from the low-cost to the high-cost location. If this line has a limit, then in periods of high demand not all the power that could be generated in the low-cost region could be used, and some of the inexpensive plants would be "constrained off." In this case, the demand would be met by higher-cost plants that absent the constraint would not run, but due to transmission congestion would be then "constrained on." The marginal cost in the two locations differs because of transmission congestion. The marginal cost of power at the low-cost location is no greater than the cost of the least expensive constrained-off plant; otherwise the plant would run. Similarly, the marginal cost at the high-cost location is no less than the cost of the most expensive constrained-on plant; otherwise the plant would not be in use. The difference between these two costs, net of marginal losses, is the congestion rental.

This congested-induced, marginal-cost difference can be as large as the cost of the generation in the unconstrained case. If an inexpensive coal plant is constrained off and an oil plant, which costs more than twice as much to run, is constrained on, the difference in marginal costs by region is greater than the cost of energy at the coal plant. This result does not depend in any way on the use of a simple case with a single line and two locations. In a real network the interactions are more complicated—with loop flow and multiple contingencies confronting thermal limits on lines or voltage limits on buses—but the result is the same. It is easy to construct examples where congestion in the transmission grid leads to marginal costs that differ by more than 100 percent across different locations.

If there is transmission congestion, therefore, the short-run market model and determination of marginal costs must include the effects of the constraints. This extension presents no difficulty in principle. The only impact is that the market now includes a set of prices, one for each location. Economic dispatch would still be the least-cost equilibrium subject to the security constraints. Generators would still bid as before, with the bid understood to be the minimum acceptable price at their location. Customers would bid also, with dispatchable demand and the bid setting the maximum price that would be paid at the customer's location. The security-constrained economic dispatch process would produce the corresponding prices at each location, incorporating the combined effect of generation, losses, and congestion. In terms of their own supply and demand, everyone would see a single price, which is the SRMC price of power at their location. If a transmission price is necessary, the natural definition of transmission is supplying power at one location and using it at another. The corresponding transmission price would be the difference between the prices at the two locations.

This same framework lends itself easily to accounting extensions to explicitly include bilateral transactions. Here market participants prefer to schedule point-to-point transmission rather than explicitly buy and sell through the spot market. The bilateral schedules would be provided to the system operator. Those not scheduled would bid into the pool-based spot market. This is often described as the "residual pool" or "net pool" approach. For market participants who wish to schedule transmission between two locations, the opportunity cost of the transmission is just this transmission price of the difference between spot prices at the two locations. This short-run transmission usage pricing, therefore, is efficient and nondiscriminatory. In addition, the same principles could apply in a multisettlement framework, with day-ahead scheduling and real-time dispatch. These extensions could be important in practice, but would not fundamentally change the outline of the structure of electricity markets.

This short-run competitive market with bidding and centralized dispatch is consistent with economic dispatch. The locational prices define the true and full opportunity cost in the short run. Each generator and each customer sees a single price for the half hour, and the prices vary over half hours to reflect changing supply and demand conditions. All the complexities of the power supply grid and network interactions are subsumed under the economic dispatch and calculation of the locational SRMC prices. These are the only prices needed, and payments for short-term energy are the only payments that apply in the short run, with administrative overhead covered by rents on losses or, if necessary, a negligible markup applied to all power. The system operator coordinates the dispatch and provides the information for settlement payments, with regulatory oversight to guarantee comparable service through open access to the spot market run by the system operator through a bid-based economic dispatch.

With efficient pricing, users have the incentive to respond to the requirements of reliable operation. Absent such price incentives, choice would need to be curtailed and the market limited, in order to give the system operator enough control to counteract the perverse incentives that would be created by prices that did not reflect the marginal costs of dispatch.

Long-run Market Contracts

With changing supply and demand conditions, generators and customers will see fluctuations in short-run prices. When demand is high, more expensive generation will be employed, raising the equilibrium market prices. When transmission constraints bind, congestion costs will change prices at different locations.

Even without transmission congestion constraints, the spot-market price can be volatile. This volatility in prices presents its own risks for both generators and customers, and there will be a natural interest in long-term mechanisms, such as contracts, to mitigate or share this risk.

Traditionally the notion of a long-term contract carries with it the assumption that customers and generators can make an agreement to trade a certain amount of power at a certain price. The implicit assumption is that a specific generator will run to satisfy the demand of a specific customer. To the extent that the customer's needs change, the customer might sell the contract in a secondary market, and so, too, for the generator. Efficient operation of the secondary market would guarantee equilibrium, and everyone would face the true opportunity cost at the margin.

However, this notion of specific performance stands at odds with the operation of the short-run market for electricity. To achieve an efficient economic dispatch in the short run, the dispatcher must have freedom in responding to the bids to decide which plants run and which are idle, independent of the provisions of long-term contracts. And with the complex network interactions, it is impossible to identify which generator is serving which customer. All generation is providing power into the grid, and all customers are taking power out of the grid. In a competitive market, it is not even in the interest of the generators or the customers to restrict their dispatch and forgo the benefits of the most economic use of the available generation. The short-term dispatch decisions by the system operator are made independent of and without any recognition of any long-term contracts. In this way, electricity is not like other commodities.

This dictate of the physical laws governing power flow on the transmission grid does not preclude long-term contracts, but it does change the essential character of the contracts. Rather than controlling the dispatch and the short-run market, long-term contracts focus on the problem of price volatility and provide a price hedge not by managing the flow of power but by managing the flow of money. The short-run prices provide the right incentives for generation and consumption, but create a need to hedge the price changes.

Consider first the case of no transmission congestion. In this circumstance, except for the small effect of transmission losses, it is possible to treat all production and consumption as at the same location. Here the natural arrangement is to contract for differences against the equilibrium price in the market. A customer and a generator agree on an average price for a fixed quantity, say 100 MW at five cents. On the half hour, if the spot price is six cents, the customer buys power from the coordinated spot market at six cents and the generators sells power for six cents. Under the contract, the generator owes the customer one cent for each of the 100 MW over the half hour. In the reverse case, with the spot price at three cents, the customer pays three cents to the system operator, which in turn pays three cents to the generator, but now the customer owes the generator two cents for each of the 100 MW over the half hour.

In effect, the generator and the customer have a long-term contract for 100 MW at five cents. The contract requires no direct interaction with the system operator other than for the continuing short-run market transactions. But through the interaction with the system operator, the situation is even better than with a long-run contract between a specific generator and a specific customer. For now, if the customer demand is above or below 100 MW, there is a ready and an automatic secondary market, namely the coordinated spot market, where extra power is purchased or sold at the spot price. Similarly for the generator, there is an automatic market for surplus power or backup supplies without the cost and problems of a large number of repeated short-run bilateral negotiations with other generators. And if the customer really consumes 100 MW, and the generator really produces the 100 MW, the arithmetic guarantees that the average price is still five cents. Furthermore, with the contract fixed at 100 MW, rather than the amount actually produced or consumed, the long-run average price is guaranteed without disturbing any of the short-run incentives at the margin. Hence the long-run contract is compatible with the short-run market.

In the presence of transmission congestion, the generation contract is necessary but not sufficient to provide the necessary long-term price hedge. A bilateral arrangement between a customer and a generator can capture the effect of aggregate price movements in the market. However, transmission congestion can produce significant movements in price that are different depending on location. If the customer is located far from the generator, transmission congestion might confront the customer with a high locational price and leave the generator with a low locational price. Now the generator alone cannot provide the natural back-to-back hedge on fluctuations of the short-run market price. Something more would be needed.

Transmission congestion in the short-run market raises another related and significant matter for the system operator. In the presence of congestion, revenues collected from customers will substantially exceed the payments to generators. The congestion rent accrues because of constraints in the transmission grid. At a minimum, this congestion rent revenue will be a highly volatile source of payment to the system operator. At worst, if the system operator keeps the congestion revenue, incentives arise to manipulate dispatch and prevent grid expansion in order to generate even greater congestion rentals. System operation is a natural monopoly, and the operator could distort both dispatch and expansion. If the system operator retains the benefits from congestion rentals, this incentive would work contrary to the goal of an efficient, competitive electricity market.

A convenient solution to both problems—providing a price hedge against locational congestion differentials and removing the adverse incentive for the system operator—is to redistribute the congestion revenue through a system of long-run transmission congestion contracts operating in parallel with the long-run generation contracts. As with generation, it is not possible to operate an efficient short-run market that includes transmission of specific power to specific customers. Once again, however, it is possible to arrange a transmission congestion contract that provides compensation for differences in prices, in this case for differences in the congestion costs between different locations across the network.

The transmission congestion contract for compensation would exist for a particular quantity between two locations. The generator in the example above might obtain a transmission congestion contract for 100 MW between the generator's location and the customer's location. The right provided by the contract would not be for specific movement of power but rather for payment of the congestion rental. Hence, if a transmission constraint caused prices to rise to six cents at the customer's location, but remain at five cents at the generator's location, the one-cent difference would be the congestion rental. The customer would pay the system operator six cents for the power. The system operator would in turn pay the generator five cents for the power supplied in the short-run market. As the holder of the transmission congestion contract, the generator would receive one cent for each of the 100 MW covered under the transmission congestion contract. This revenue would allow the generator to pay the difference under the generation contract so that the net cost to the customer is five cents as agreed in the bilateral power contract. Without the transmission congestion contract, the generator would have no revenue to compensate the customer for the difference in the prices at their two locations. The transmission congestion contract completes the package.

When only the single generator and customer are involved, this sequence of exchanges under the two types of contracts may seem unnecessary. However, in a real network with many participants, the process is far less obvious. There will be many possible transmission combinations between different locations. There is no single definition of transmission grid capacity, and it is only meaningful to ask if the configuration of aggregated transmission flows is feasible. However, the net result would be the same. Short-run incentives at the margin would follow the incentives of short-run opportunity costs, and long-run contracts would operate to provide price hedges against specific quantities. The system operator coordinates the short-run market to provide economic dispatch. The system operator collects and pays according to the short-run marginal price at each location, and the system operator distributes the congestion rentals to the holders of transmission congestion contracts. Generators and customers make separate bilateral arrangements for generation contracts. Unlike with the generation contracts, the system operator participation in coordinating administration of the transmission congestion contracts is necessary because of the network interactions, which make it impossible to link specific customers paying congestion costs with specific customers receiving congestion compensation. Still, congestion prices will be highly variable and load dependent. Only the system operator will have the necessary information to determine these changing prices, but the information will be readily available embedded in all the spot-market locational prices. The transmission congestion contracts define payment obligations that guarantee protection from changes in the congestion rentals.

Were it possible to define usage of the transmission system in terms of physical rights, it would be desirable that these rights have two features. First, they could not be withheld from the market to prevent others from using the existing transmission grid. Second, they would be perfectly tradable in a secondary market that would support full reconfiguration of the patterns of network use at no transaction cost. This is impossible with any known system of physical transmission rights that parcel up the transmission grid. However, in a competitive electricity market with a bid-based, security-constrained economic dispatch, transmission congestion contracts are equivalent to just such perfectly tradable transmission rights. Hence we can describe transmission congestion contracts either as financial contracts for congestion rents or as perfectly tradable physical transmission rights.

If the transmission congestion contracts have been fully allocated, then the system operator will be simply a conduit for the distribution of the congestion rentals. The operator would no longer have an incentive to increase congestion rentals: any increase in congestion payments would flow only to the holders of the transmission congestion contracts. The problem of supervising the dispatch monopoly would be greatly reduced. And through a combination of generation contracts and transmission congestion contracts, participants in the electricity market could arrange price hedges that would provide the economic equivalent of a long-term contract for specific power delivered to a specific customer.

Further to the application of these ideas, locational marginal cost pricing lends itself to a natural decomposition. For example, even with loops in a network, market information could be transformed easily into a hub-and-spoke framework with locational price differences on a spoke defining the cost of moving to and from the local hub, and then between hubs. This would simplify without distorting the locational prices. A contract network could develop that would be different from the real network without affecting the meaning or interpretation of the locational prices.

With the market hubs, the participants would see the simplification of having a few hubs that capture most of the price differences of long-distance transmission. Contracts could develop relative to the hubs. The rest of the sometimes important difference in locational prices would appear in the cost of moving power to and from the local hub. Commercial connections in the network could follow a configuration convenient for contracting and trading. The separation of physical and financial flows would allow this flexibility.

The creation or elimination of hubs would require no intervention by regulators or the system operator. New hubs could arise as the market requires, or disappear when not important. A hub is simply a special node within a zone. The system operator still would work with the locational prices, but the market would decide on the degree of simplification needed. There would be locational prices, and this would avoid the substantial incentive problems of averaging prices.

LONG-TERM MARKET INVESTMENT

In the case of investment in new generating plants or consuming facilities, the process is straightforward. Under the competitive assumption, no single generator or customer is a large part of the market, there are no significant economies of scale, and there are no barriers to entry. Generators or customers can connect to the transmission grid at any point, subject only to technical requirements defining the physical standards for hookup. The investor takes all the business risk of generating or consuming power at an acceptable price. If the generator wants price certainty, then new generation contracts can be struck between a willing buyer and a willing seller.

If either party to the contract expects significant transmission congestion, then a transmission congestion contract would be indicated in any case that would otherwise benefit from point-to-point physical rights were they available. If transmission congestion contracts are for sale between the two points, then a contract can be obtained from the holder(s) of existing rights. Or new investment can create new capacity that would support additional transmission congestion contracts. The system operator would participate in the process only to verify that the newly created transmission congestion contracts would be feasible and consistent with the obligation to preserve any existing set of transmission congestion contracts on the existing grid. Hence, incremental investments in the grid would be possible anywhere without requiring that everyone connected to the grid participate in the negotiations or agree to the allocation of the new transmission congestion contracts.

This happy resolution of the puzzle of transmission expansion and pricing through voluntary market forces alone is subject to at least two other important caveats. First, there still may be market failures even with the definition of a workable set of equivalent property rights. For example, with many small-market participants, each benefiting a little from a large transmission investment, the temptation to free-ride on the economies of scale and scope may create a kind of prisoner's dilemma. Everyone would be better off sharing in the investment, but the temptation to free-ride and avoid paying for the expense may overcome any ability to form a consortium or negotiate a contract. It may be that the investment could not go forward in a timely manner, at the right scale, or at all, without some entity that can mandate payment of the costs.[5] In this case, however, the task should be simplified by the ability to simultaneously allocate the benefits in the form of a share of the transmission congestion contracts. The market could take care of many, perhaps most, investments, and the regulatory option would be easier to implement when needed.

A related problem could appear in the circumstances where the pattern of transmission use was so uncertain and the network so interconnected that no set of point-to-point rights would be capable of capturing enough of the economic benefits of grid investment. This would be true, of course, for both physical rights, were they possible, and the transmission congestion contracts. In effect, there would be significant economies of scope in transmission investment that would go well beyond the benefits of any reasonable patterns of point-to-point rights. If the benefits could not be assigned, then the market-based investments would not follow.

Second, operation of voluntary market forces would have little sway in the allocation of the costs for an existing transmission grid that already provides open access. The costs are sunk, and typically the sunk costs of the wires exceed the transmission congestion opportunity costs of using the grid. This is due, in large part, to the effects of economies of scale. Hence, given the choice of paying the sunk costs but avoiding the congestion costs, versus avoiding the sunk costs while using the system and paying the continuing cost of congestion, most users would prefer the latter. If the sunk costs are to be recovered in prospective payments, therefore, there must be some form of requirement to pay these costs as a condition for using the grid. The resulting access charges would be the functional equivalent of the contract payments for new investment.

The need for access charges would not dictate the form of the charges. From an efficiency perspective, the preference would be to divorce the charge from usage levels, such as through a meter connection charge. In practice, the access charge might be collected proportional to usage, but in this case it should be at the last point of connection in order to minimize perverse incentives for inefficient bypass to avoid paying for sunk costs.

Assignment of the access charges could be simplified by the simultaneous allocation of the benefits of the existing grid through the award of transmission congestion contracts. One approach might be to award the initial ownership of the contract along with a long-term obligation to pay the access charges. Another possibility would be to auction the transmission congestion contracts and apply the revenues to reduce the required payment for the transmission grid. The remaining costs would be collected as access charges for all users of the grid. In effect, these access charges would pay for the assets of the grid in a system allowing open access for full use of the system. Everyone who used the system would pay the transmission usage charge for congestion, the second part of a two-part tariff, and the holders of transmission congestion contracts would have the transmission hedge.

Grid expansion and pricing would continue to present a need for regulatory oversight, but the existence of workable transmission congestion contracts would substantially simplify transmission investment decisions. The users of the system who are buying and selling electricity without a complete hedge through transmission congestion contracts would face the short-term market clearing price. In the face of transmission congestion, the locational prices provide the proper incentive for investment in transmission facilities. Investments should be made when justified by the savings in congestion costs. Those who are prepared to make the investment would obtain the associated transmission congestion contracts. The role of regulators, therefore, would be to review requests for transmission expansion, examine the compatibility with the companion request for new transmission contracts, and ensure an open process for all to join in developing combined transmission investments recognizing the interactions in the network.

The regulator would be responsible for enforcing a requirement for existing transmission facility owners to support expansions and reinforcements at a traditional regulated cost that recovered the incremental investment, and then to assign the corresponding transmission contracts. If no coalition of grid users was able to agree to pay for a grid expansion that appears to be beneficial for the system as a whole, any interested party could propose a project and an allocation of its costs among those grid users who would benefit. Regulatory procedures, similar to those used now, would determine whether the project should go forward and how its costs should be allocated to those expected to benefit from the effect on future locational spot-market prices, with the payers granted rights to compensation to assure that future congestion does not rob them of the benefits they are paying for.[6] This would also be the place to take up the related question of providing adequate incentives, and the form of incentive regulation, for transmission providers who provide connection to the existing grid and construction of expansions that would not take place under the market-based framework (see figure 7.2).

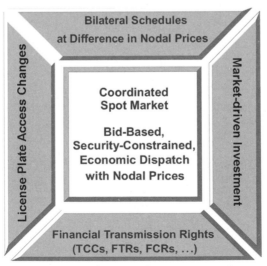

Figure 7.2.   The RTO-NOPR Contains a Consistent Framework

The transmission congestion contracts, once created, would no longer need any special regulation. Although investments in the transmission grid would be lumpy and would require the cooperation of the owners of existing facilities, the transmission congestion contracts would be divisible and freely tradable in a secondary market. This secondary market would provide a ready source of transmission hedges that would serve as an alternative to system expansion. The price of the transmission contracts should never rise above the long-term expected congestion opportunity costs or the cost of incremental expansion of the grid.

Regional Transmission Organizations

The basic components of this competitive market structure appear in the proposed (Editor's note: as of 1999) requirements for Regional Transmission Organizations.[7] The key element is in the recognition of the importance of a coordinated spot market. In the RTO NOPR (Notice of Proposed Rulemaking) this appears as the balancing market, which is equivalent to the "net pool" arrangement outlined above. In particular, Federal Energy Regulatory Commission (FERC) recognizes that "[r]eal-time balancing is usually achieved through the direct control of select generators (and, in some cases, loads) who increase or decrease their output (or consumption in the case of loads) in response to instructions from the system operator."[8] To be consistent with the competitive market, it is essential that this be through a bid-based security-constrained economic dispatch: "Proposals should . . . ensure that the generators that are dispatched in the presence of transmission constraints must be those that can serve system loads at least cost, and limited transmission capacity should be used by market participants that value that use most highly."[9]

Given the availability of this coordinated spot market and these efficient locational prices, market participants could schedule bilateral transactions or rely on trade through the spot market. The differences in locational prices would define the opportunity costs of transmission, giving rise to the creation of financial transmission rights.[10] Payment for the existing grid would appear in part as access charges, including the use of the "license plate" approach with region-specific access charges.[11]

This same basic system has been in operation since 1998 in the Pennsylvania-New Jersey-Maryland (PJM) Interconnection, where the financial transmission rights are labeled as fixed transmission rights (FTRs).[12] Similar systems exist in the New York and New England ISOs. There are slight differences in all these approaches, but they stand on the bedrock of a coordinated spot market, implemented through a bid-based, security-constrained economic dispatch with locational prices. This supports a high degree of choice by market participants and is the only known model that provides these benefits in a framework to support competitive electricity markets.

NETWORK EXTERNALITIES AND INVESTMENT INCENTIVES

Complete reliance on market incentives for transmission investment would be unlikely as a practical matter and is subject to a number of theoretical challenges. Market institutions can support market-based transmission investments. The tradeoff will be between imperfect markets and imperfect regulation. The existence of a viable system of long-term transmission congestion contracts changes the balance, however, to rely more on market forces. Here we address the challenges to market-based investments that arise because of network externalities and economies of scale and scope.

Allocating Transmission Benefits and Costs

The economies of scale and scope in transmission construction present a number of issues in long-run transmission pricing. Transmission investments and upgrades tend to come in large increments that can have a significant effect on market prices. The facilities are capital intensive, require long investment lead times, and are long lived. Any change in the configuration of the transmission grid can alter the extensive network interactions, thereby affecting virtually all users of the transmission grid. Furthermore, most of the costs for the existing transmission system are sunk. Hence, there is a marked difference between the problems of determining the allocation of the costs for the existing system and the task of defining the allocation for system expansion.

The incentive for transmission investment would come from those who did not already have transmission congestion contracts. These market participants would be relying on the spot market. When the anticipated congestion cost and losses become large enough, there would be an economic motivation for investing in transmission. However, the motivation for the investment assumes, at a minimum, that those making the investment would then reduce the cost of the losses and avoid the corresponding congestion payments.

The discussion of the economics of competitive electricity markets emphasizes the importance of defining and allocating benefits of the transmission grid. If there is no method to allocate the benefits of transmission, the strong network interactions exacerbate the difficulty of allocating costs. Without some form of transmission rights, there could be substantial incentives to shift costs to others (i.e., free-riding), it would be difficult to define a standard of performance for transmission providers, and the normal operation of market principles would founder because of the lack of property rights. By contrast, where it is possible to define the equivalent of property rights, free-riding and these difficult allocation problems could be substantially reduced, if not eliminated.

Consider the allocation of the costs for short radial connections from a plant to a nearby transmission grid. In the simplest of cases, when connecting a single generator or a single customer, the benefits of the connection seem clear. The benefits would accrue to the facility connected to the grid, and the costs would naturally be allocated to the same facility. The close match between benefits and costs would allow for investment charges that would be consistent with operation of a market. It is for this reason that transmission pricing often treats such investments separately.

When we turn from simple radial connections to the main transmission grid, however, the obvious allocation of costs and benefits is no longer available. This is closely related to the complicating fact of network interactions that make it difficult or impossible to define the total capacity of a network. The simple reality is that in any sufficiently interconnected network, it is not possible to define the capacity of the network without also defining the pattern of usage. Except for truly radial systems, transmission congestion can change the capacity of the grid depending on the configuration of load and generation.

This difficulty in defining the capacity of the grid without specifying the pattern of use does not mean that no definition of capacity or property rights would be available. For example, it is always possible to test to determine if a particular pattern of usage of the transmission grid would be within the security-constrained dispatch limits dictated by the requirements of reliable operation. The system operator makes just such judgments on a regular basis, with an affirmative conclusion implicit in every dispatch. Hence, it would be legitimate to define any feasible pattern of system use as within the capacity of the system, and the associated input levels could be defined as transmission rights. With these hypothetical property rights, we would be closer to a method for allocating the benefits of the transmission grid to support the allocation of the costs.

One difficulty of this approach in terms of physical property rights would be in the constant need to redefine and reconfigure transmission usage to conform to the actual load patterns. In effect, we would like those who have paid for transmission property rights to be able to sell those rights in a secondary market to support the actual dispatch. Were this possible, many of the most perverse incentives and logical puzzles in transmission pricing would be removed. With well-defined, tradable property rights, the transmission regime could operate more according to market principles. However, a literal system of physical rights traded to match actual use would be impractical as a means to control system dispatch.

This is where the transmission congestion contracts fit into the long-run picture. A transmission congestion contract provides payment equal to the congestion price differential between two locations. In a competitive market setting, these financial contracts are equivalent to perfectly tradable physical transmission rights. Hence, these transmission congestion contracts can be the embodiment of the benefits of transmission investment. Ownership of the transmission congestion contract can be defined as the benefit that comes with paying for transmission investment. The transmission congestion contract provides a perfect hedge against changes in the transmission usage charges arising in the form of congestion costs. The transmission congestion contract internalizes the complicated network interactions, and allows the market to operate with simple point-to-point arrangements that act like property rights.

If there were no market power present, the value of the opportunity costs under the transmission congestion contracts would define the opportunity costs of any reduction of system capacity. This would set a framework for any payment obligations by the transmission provider. In the presence of market power in generation or load, including the important special case of a radial connection for a single entity, the transmission congestion contract obligation could be adapted to specify obligations of the transmission provider. For example, in case of the loss of a radial line, there would be no observed market clearing price at the plant to define the opportunity cost of the reduced capacity caused by a line outage. However, in the context of a spot market, a pattern of bids from previous periods might establish a workable estimate of the incremental cost of the plant. The difference relative to the grid price at the point of connection could be used as the estimate of the opportunity cost in the rare event of line outage. In this case, the opportunity cost estimate would serve as the implied obligation under the transmission congestion contract.

Inefficient Transmission Expansion

In the case of electricity the focus of inefficient expansion is on reducing the capacity of the network, not expanding too much. It is well known that different pricing regimes can create incentives for "uneconomic" transmission investment in the sense of reducing total welfare by reducing the effective capacity of the system.[13] The simple example of a radial transmission line between two points would illustrate the point. Suppose that the line connects a low-cost region to a high-cost region. Under market pricing rules, generators in the high-cost region would benefit if the capacity of the line were reduced. In an electric grid, a conceptually simple way to reduce the transfer capacity would be to add a weak parallel line between the two regions. Power flow would now split between the two parallel lines. Flow on the original parallel line would then have to be reduced to keep the parallel flow on the new line below its low capacity. It might be worth it to the high-cost generators to build such a line, lower transfers, and raise prices in the high-cost region. Overall economic efficiency would be reduced, but the profit to the high-cost generator could justify the expenditure for the transmission investment.

Most of the examples of such perverse transmission investment incentives rely on this property of parallel flows. This simple perverse example would be easy to dispose of in practice. The regulator might prohibit such degradations of capacity. However, in a real network, with complex interactions, the same effect may be present but less obvious. The analysis to reveal a truly uneconomic investment might be both more difficult and controversial. Hence, we cannot rule out the possibility that wealth transfers that dominate efficiency losses would influence real investment incentives. In theory, decentralized market-based investments could be inefficient.

It would be desirable, therefore, to have a system of property rights that reduced or eliminated the incentives for such perverse investments. In the case of a contract network as described above, the investment incentives would be affected by the existence of the transmission congestion contracts. As demonstrated by Bushnell and Stoft, under certain restrictions, the existence of transmission congestion contracts may even fully internalize the efficiency effects of investment decisions.[14]

The allocation of transmission congestion contracts under a feasibility rule arises naturally from the interpretation of the transmission congestion contracts as described above. It is clear that simultaneous feasibility of the transmission congestion contracts is necessary to guarantee that the revenues collected by the system operator are sufficient to pay the obligations under the transmission congestion contracts. Furthermore, under certain minimal conditions, the same feasibility condition would be sufficient to guarantee that the revenues from market equilibrium in the spot market would be sufficient to guarantee this revenue adequacy.[15] However, in addition, the allocation of transmission congestion contracts under the same feasibility rule mitigates incentives for inefficient transmission investment.

Accommodating Economies of Scale

The electricity industry with transmission investment is not the only industry that can face significant economies of scale in investment. Very large electric transmission expansions might be an extreme case, but the principle applies elsewhere in expansion of aluminum production capacity, steel plants, hotels, and so on. In these other markets, we accept and accommodate the effects of economies of scale. The result is a "second-best" outcome, narrowly defined. But we accept the benefit of the compromise given the reality of an imperfect market, and do not reject the results of this market in the absence of a better solution through regulation. If we have a better approach through regulation, we should take it. But if not, we should enjoy the second-best outcome of the market. The same approach could be extended to electricity transmission investment.

One accommodation would take the form of allowing market-based investments that are scaled to be small enough to have a minimal impact on market prices. Hence, the investor would be able to profit from the remaining price differences in the spot market, while enjoying the increased throughput provided by the transmission expansion. A careful engineering analysis might show that a larger project would be more "efficient," but in the absence of volunteers to pay for the expansion, it would be better to have a smaller expansion than no expansion at all. The principal public policy requirement would be to ensure that the project did not fully occupy some unique resource, such as a right of way, that would no longer be available once the investment went forward. As long as the investment would be subject to further entry and competition, the second-best argument would be to support the market-based investment and its contribution to dynamic efficiency.

More controversial would be to encourage market-based investment at a greater scale but to allow a period of time where the investment would be operated at reduced capacity. Here there would be an analogy to patents on inventions. With a patent, an inventor can restrict use of an invention that could easily be used much more widely, but only at the cost of dissipating the profits. The withholding limit applies for a period of time, judged sufficient to stimulate the effort of invention. Similarly, the transmission investor would enjoy a period of time where the profits from the resulting locational price differences would be sufficient to justify the investment, even though operation at full capacity would make the investment unprofitable.

There would be no technical difficulty in operating these incremental investments at less than the full name-plate capacity. All that would be required would be the cooperation of the system operator in specifying the limits to apply in the security-constrained economic dispatch. In the very nature of the security-constrained dispatch, with multiple contingency constraints, the limit is monitored based on a calculation, not just from measuring the power flow over a line. The calculation could just as well be based on the commercial limit for the investment as on the physical limit.

The idea of allowing withholding of actual physical capacity seems at odds with the policy of open access and full utilization of the existing grid. However, on closer inspection there is no fundamental contradiction. The market-based investment would not be like the regulated investment. In particular, there would be no rate base protection or inclusion of the investment cost in mandatory access charges. For the capacity associated with regulated investments, there would be no withholding.[16] But for market-based investments, we recognize that the investment might not be made if the profit incentive were removed by requiring optimal exploitation of economies of scale and immediate utilization of the full capacity. The investor would bear the risk of the market-based investment, but would require the reward of charging a price to justify the investment.

As long as the investment did not foreclose competitive entry, the withholding should not be seen as the deleterious exercise of market power. Rather, the temporary withholding could be seen as part of the process of dynamic adjustment in the market, with the profit incentive originating in the scarcity rents. It is precisely the pursuit of these scarcity rents that motivates entry and the investment. Eventually, further entry would dissipate the scarcity rents or the period of "patent" protection would expire. In the interim, the expansion would provide direct incremental benefits to the energy market, and the cushion between commercial and physical capacity would provide an improvement in overall reliability.

The logic here would be the same as applied to the other industries. We do not expect investors to build hotels with the promise to lease out rooms at the marginal cost of maintenance until the hotel is full, when the scarcity rent applies. So too here we would not expect the investor in market-based transmission to expect to incur a loss.

Furthermore, it would be difficult to make an operational distinction between smaller-scale investments with the commercial capacity equal to the physical capacity, and those larger-scale investments where there was an acknowledged gap. Measurement of physical capacity is not without some judgment, so the issue could be clouded. Furthermore, at some cost it would be possible to downgrade the physical capacity to match the commercial requirements. It would be more efficient, therefore, to allow operation of the dispatch where new investment could be defined as an increment to capacity that would be acceptable to the investor, as long as the physical capacity at least equaled the commercial recognition of the capacity.

This dynamic investment process stands behind the expansion rules in Australia.[17] The basic idea was to provide an incentive for market-based transmission investment. In the case of Australia, there is a safe harbor provision that allows the investor to effectively control the capacity of a new transmission line. In effect, this allows the investor to limit the use of the line whenever the locational price difference is too low. To the extent that this maintains higher price differences, the withholding is justified as the necessary cost of reaping the benefits of the expansion. As a result, more power flows and congestion costs are reduced, compared to the case without the investment.

In the case of the Australian safe harbor provision, application is limited to the case of "controllable" lines. This simplifies the model substantially, because it mitigates or eliminates the network externalities. Furthermore, the use of a controllable line approximately restores the linkage between the identifiable increment in system capacity and the actual flow over the line. In practice, therefore, this allows for a payment method that is equivalent to charging a fee for using the line. In this Australian case, the limitation to a controllable line is necessary because Australia does not have the foundation of a bid-based, security-constrained economic dispatch with fully locational prices. Hence, there are unaccounted-for network interactions, and the counterparts to transmission congestion costs have not been implemented. For a controllable line the economic model is still well defined by the flows over the line. However, if Australia were to adopt full locational pricing, which would be advantageous for other reasons, it would be possible to expand the safe harbor provisions to transmission investments other than physically controllable lines. This would encompass all the possible investments, including new free-flowing lines, capacitor and transformer upgrades, and so on.

In the event, as soon as Australia created this small niche for a market-based transmission investment, construction began on just such an expansion. This is the 180 MW Direct Link project connecting the regional electricity markets in Queensland and New South Wales.[18] This is the cleanest demonstration in support of the argument that market-based investment in transmission can and will be made, given the right institutional framework. The project is a pure transmission investment, and the investor is taking the risk while expecting to arbitrage the price differences between the markets. The project is a controllable underground DC line, which provides a number of special features. But the principle could be applied to support other types of market-based investments.

Absent artificial barriers to entry, the market could discipline this process. Furthermore, this institutional innovation would not foreclose the regulatory alternative. The investor could always agree to make available the full capacity in exchange for a sufficient payment to cover the cost of the investment. But now the regulator, rather than the customer, would be making the deal. And the collection of the fixed charges would be part of the mandatory regime of access charges.

CONCLUSION

Market-based transmission investments can play a significant role in a competitive electricity market, if given the opportunity. However, market institutions must be designed to support the competitive market and the transmission investments. A short-term electricity market coordinated by a system operator through a bid-based, security-constrained economic dispatch provides a foundation for building a system that includes tradable transmission property rights in the form of transmission congestion contracts. Coordination through the system operator is unavoidable, and spot-market locational prices define the opportunity costs of transmission that would determine the market value of the transmission rights without requiring physical trading and without restricting the actual use of the system. In this setting, these transmission congestion contracts are equivalent to perfectly tradable physical rights. Hence, this organization of the market defines a context where it would be possible to rely more on market forces, partly if not completely, to drive transmission expansion. In the case of market failure, transmission investment would still require some regulatory supervision, but this case would be simplified through the use of transmission congestion contracts.

NOTES

1. This chapter draws on work for the Harvard Electricity Policy Group and the Harvard-Japan Project on Energy and the Environment. The author is or has been a consultant on electric market reform and transmission issues for American National Power, British National Grid Company, GPU Inc. (and the Supporting Companies of PJM), GPU PowerNet Pty Ltd, Duquesne Light Company, Electricity Corporation of New Zealand, National Independent Energy Producers, New England Power Company, New York Power Pool, New York Utilities Collaborative, Niagara Mohawk Corporation, PJM Office of Interconnection, San Diego Gas & Electric Corporation, TransÉnergie, Transpower of New Zealand, Westbrook Power, Williams Energy Group, and Wisconsin Electric Power Company. The views presented here are not necessarily attributable to any of those mentioned, and any remaining errors are solely the responsibility of the author.

2. David M. Newbery, "Privatization and Liberalization of Network Utilities," *European Economic Review*, Vol. 41, 1997, pp. 357–83.

3. James B. Bushnell and Steven E. Stoft, "Grid Investment: Can a Market Do the Job?" *The Electricity Journal*, January–February, 1996, pp. 74–79. Eric C. Woychik, "Competition in Transmission: It's Coming Sooner or Later," *The Electricity Journal*, June 1996, pp. 46–58.

4. Here we assume that other externalities, such as environmental effects, have been internalized.

5. This situation appears to be what is described often as investments for reliability. However, with price-responsive demand and security-constrained economic dispatch, there is in principle no difference in reliability investments and economic investments. The only difference created by the investment would be in the economic benefits of the actual dispatch.

6. D. Garber, W. Hogan, and L. Ruff, "An Efficient Electricity Market: Using a Pool to Support *Real* Competition," *The Electricity Journal*, Vol. 7, No. 7, September 1994, pp. 48–60.

7. Federal Energy Regulatory Commission, "Regional Transmission Organizations," Notice of Proposed Rulemaking, Docket No. RM99-32-000, Washington, DC, May 13, 1999. (RTO NOPR)

8. RTO NOPR, p. 175.

9. RTO NOPR, p. 198.

10. RTO NOPR, p. 166.

11. RTO NOPR, p. 13.

12. PJM Interconnection, L.L.C. The PJM system began operation with FTRs and full locational pricing as of April 1, 1998. Details can be obtained from the website at www.pjm.com.

13. For a review of the literature on these incentives, see James B. Bushnell and Steven E. Stoft, "Electric Grid Investment Under a Contract Network Regime," *Journal of Regulatory Economics*, Vol. 10, 1996, pp. 61–79.

14. James B. Bushnell and Steven E. Stoft, "Improving Private Incentives for Electric Grid Investment," *Resource and Energy Economics*, Vol. 19, 1997, pp. 85–108. See also Bushnell and Stoft, "Electric Grid Investment Under a Contract Network Regime." For a related analysis in the context of generator market power, see Yves Smeers and Wei Jing-Yuan, "Transmission Contracts May also Hinder Detrimental Network Investments in Oligopolistic Electricity Markets," *CORE*, October 28, 1997.

15. Scott M. Harvey, William W. Hogan, and Susan L. Pope, "Transmission Capacity Reservations and Transmission Congestion Contracts," Center for Business and Government, Harvard University, June 6, 1996 (Revised March 8, 1997).

16. This is for regulated investments under the model envisioned here. Under the contract path model, the variable wheeling charges independent of congestion amount to withholding through price.

17. National Electricity Code Administrator, "Entrepreneurial Interconnectors: Safe Harbour Provisions," *Transmission and Distribution Pricing Review*, Australia, November 1998.

18. The Direct Link project is a merchant transmission line in Australia developed by TransÉnergie. Further details on the Direct Link project can be found on the TransÉnergie website at www.transenergie.com.au. The author has been advising TransÉnergie U.S. Ltd. in considering merchant transmission investments in the United States.

8

Preventing Monopoly or Discouraging Competition? The Perils of Price-Cost Tests for Market Power in Electricity

Timothy J. Brennan

Concerns have been expressed around the world that newly opened wholesale electricity markets have failed to be sufficiently competitive.[1] The competitiveness of electricity markets in the United Kingdom has been questioned almost since their inception.[2] The California electricity market "meltdown" in the summer of 2000 brought with it numerous accusations and analyses of the role that inadequate competition played in creating price spikes and destabilizing the market.[3] Recently, Short and Swan have produced a thoughtful study of market power in the Australian electricity sector.[4] The study is especially valuable for clear and useful displays of bidding data.

There are sound theoretical reasons for believing that electricity markets may be unusually susceptible at times to the exercise of market power, compared to markets for other goods with otherwise similar competitive characteristics (e.g., measures of market concentration). When it comes to the empirical assessment of market power, however, the approach taken in most of the analyses of market power in electricity rests on a flawed application of a standard measure of market power—the Lerner index, also known as the price-cost margin.[5] The fundamental rationale for using price-cost margins is essentially that in a competitive market, price-taking firms will supply output up to the point where the marginal cost of production just equals the market price. Therefore, a substantial difference between price and marginal cost indicates that firms are not taking price as given.

In a nutshell, the flaw in many electricity market studies is not that the price-cost margin is theoretically inappropriate, but the manner in which it has been implemented. In these studies, the proxy for "marginal cost" used to estimate price-cost margins is typically the average variable or operating cost of the "last" generator that would be dispatched to meet energy demand.[6] Let us call this the PAVC test, for "price-average variable cost."[7] As Short and Swan put it,

> [Competitive] behavior will give rise to the lowest market price that ensures that all generators are at least compensated for any short-run marginal costs incurred. Under these conditions, the market price would reflect the short run marginal cost of the most expensive generation turbine called to supply into the market.[8]

The PAVC standard for competitive pricing would imply that no generator would be built. In any market, competitive or not, even this most expensive "marginal" generator has to expect that prices will, on average, cover not just its variable costs but its fixed capital costs as well.[9] If not, it would find entry unprofitable. As we discuss below using resort hotels as an example along with electricity generation, the need to recover fixed costs can lead to prices substantially above average variable costs in peak periods. From that starting point, it is not difficult to imagine enough real-world "noise" in the form of uncertainty regarding demand and supply shocks (e.g., unanticipated hot weather, generator outages) leading to patterns of market bids similar to those found in the aforementioned studies, without necessarily indicating market power. A better method for ascertaining the extent of market power in electricity would be to focus on output, looking for generation capacity that would have been profitable to run at prevailing market prices, but was not.

Erroneous use of price-cost margins to measure market power is not merely a matter of academic interest. Such measures could lead, and perhaps already have led, regulators to prevent sales of electricity above the highest average variable cost of the generators used to provide electricity. Keeping at least some firms from earning revenues in excess of their average variable costs will encourage present suppliers to leave the market and discourage new firms, particularly those needed to provide power on-peak, from entering. Without such entry, opening electricity markets is likely to fail. Ironically, limited regulation intended to ensure competitive performance would subvert entry, leading ultimately to a reduction in such competition and perhaps a restoration of full-blown rate-of-return based price controls.

THE INTUITION

The primary context in which market power might be exercised is when the industry is facing capacity constraints. In that context, when one is trying to discover what the (short-run) competitive price would be in a market where capacity is limited, one would not compare price to the average variable cost of the marginal plant, or even the next one that might have been brought online. In a simple model where there are only two levels of demand, peak and off-peak, one would predict that the peak price over the long run would equal that highest average variable cost plus the average capacity cost of the plant.[10]

Marginal-cost measures can be appropriate, but only if one rejects using average variable costs as a proxy. A marginal plant recovers its capital costs in a competitive market because at some point the marginal costs *within the plant* are increasing. The example used here—a generation unit with constant marginal costs within the plant (not across plants) up to a capacity constraint, at which point marginal cost is effectively infinite—is an extreme example of that phenomenon. Generally, one would expect marginal cost to increase as one gets closer to "full capacity," if for no other reason than one may run a greater risk of a breakdown of the unit down the road. If so, those costs would need to be included in a Lerner index to see if prices are inappropriately high. Calculating marginal costs based on fuel prices and other average operating and maintenance costs, as is typical in these studies, does not recognize this phenomenon. At peak periods, average variable cost is not a proxy for marginal cost, accurately conceptualized and measured.

The measurement situation is even worse. Over the life of a peak plant, demand during peak periods itself will vary. Some peak hours, the demand will be high; during some it will be low, yet absent outages, its output would be constant, as illustrated in figure 8.1.

Peak-period price can thus vary with demand, even if firms are price takers. Yet because output remains fixed, any measure of cost that one would want to use would remain constant. Consequently, a measure of market power based on a relationship between price and a static proxy for marginal cost, such as average variable cost, is inherently inadequate. A measure does not work if one can get different values while the underlying phenomenon—in this case, failure to act like a price-taker—does not change. Either a price-based measure does not indicate the level of market power in such industries,

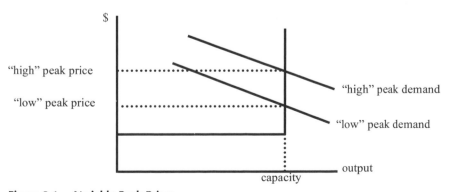

Figure 8.1.   Variable Peak Prices

or one is claiming that peak prices are somehow anticompetitive, and more so as demand rises. Neither conclusion is acceptable.

With demand varying over time, the only way to know if these prices are inappropriately high relative to "marginal cost" would be to compare the discounted present value of revenues received from electricity sales from the unit to the total construction and operation costs of unit. Even if that virtually impossible task could be carried out, it would be impossible to conclude simply on that basis that firms had been acting anticompetitively. They may simply have underestimated demand when they constructed the units in the first place, or there may be regulatory rules that limit plant construction (or expansion).

Studies of market power based on price-cost margins reflect virtually no appreciation of these concerns. Marginal cost is not the average operating cost of the most expensive natural gas plant, based on gas prices, emission permit prices (in parts of the United States)[11] and other variable costs. Unless one posits that we have an overbuilt industry, in the sense that the peak plants are destined to lose money, peak plants are going to earn capacity rents in a well-functioning, competitive market.

Perhaps the best and least fortunate example is that the Federal Energy Regulatory Commission (FERC) has used the "highest average variable costs" standard in setting its wholesale price cap, explicitly saying that it will not allow higher prices so that firms can earn capacity rents.[12] Under such a policy, in the long run no firm would build a peaking plant. Moreover, no firm would enter the industry at all, if the firm with highest operating cost is not allowed to recover its capital expense.[13]

HOTEL ROOMS

To get a feel for the flaw in the PAVC test, let us turn first to a more familiar industry—resort hotels. Imagine that in a seaside town, one can build hotels. The optimal size for a hotel is 100 rooms. Once built, it costs $50 per day to maintain a room, including cleaning, electricity, water, and predictable wear-and-tear from usage. The fixed annual capital costs for the hotel are $1,095,000 per year ($30 per day per room, for 365 days and 100 rooms). There is no relevant restriction on entry (i.e., if one thinks that one can profitably operate a 100-room hotel in this town, one can build it). To make the example simple, we assume that the firms are acting competitively (i.e., take the going room rate as given in making decisions whether to build a new hotel).

Suppose first that demand to use this resort is roughly the same all year round. In that case, hotels will enter up to the point where the price of a room is $80 per day. Of that $80, $50 covers the cost of maintaining a room—the average variable cost—while $30 of the $80 goes to cover the capital cost of the hotel. At prices above $80, more hotels would be built. If price were forecast to be below $80, say $50, no one would enter. The PAVC test would fail to predict competitive prices in the market.

Next, imagine that demand for hotel rooms at this resort town is seasonal. For three months out of the year, people really want to come to the beach. The rest of the time, demand for rooms is weak. In such a situation, a decision to build a new hotel will be predicated on filling it up during the summer season. Accordingly, the price of hotel rooms in the summer will be $170 per day; $50 of this rate is the average variable cost, and $120 is needed to cover the cost of the hotel entirely from summer occupancy.

Because every hotel gets to charge this rate during the summer, not only those hotels built to serve summer clients, they all will capture their capital costs at that time. The price of a room off-season would then be only $50. The PAVC standard would correctly predict off-peak rates, but would fail on-peak, setting them also at $50 when the competitive rate would be $170. Holding hotels to a PAVC standard would mean that none would be built to serve summer visitors to the resort. It would also imply that year-round hotels would be unable to recover their capital costs as well.

There is one case when the PAVC test might be relevant. Suppose interest in visiting this resort fell dramatically after hotels were already built (e.g., because the water was found to be unsanitary or fear from shark attacks). The hotels could not be "deconstructed," so to speak. Hotels would compete through reduced prices until the hotels already constructed were full at a rate below $170—perhaps visitors want to sit on the beach, and not go in the water—or prices fell to average variable cost, $50. Only if the market had large amounts of excess capacity, defined in terms of the long-run unprofitability of a new entrant, might we expect a PAVC standard to apply. But expecting excess capacity as a permanent feature would be unrealistic unless the tourism industry were in permanent decline.

BACK TO ELECTRICITY

Peak-load pricing principles that hold for hotels regarding peak-load pricing hold for electricity as well. We will get to some important complications, but first imagine that there is only one kind of electricity generator with 100 megawatts of capacity, with average variable costs of, say, $30 per megawatt-hour (MWh). Suppose also that of the 8760 hours in a year, demand is at peak for 450 hours, about 2 percent of the time. Finally, suppose that the fixed annualized costs of building and maintaining the generator is $7.65 million, a figure chosen to come out to $170 per MW per peak hour. (This is also about 30 percent of the total variable cost of running a plant full out.) For simplicity, again, assume that at off-peak times capacity exceeds the amount of electricity demanded at $30 per MWh.

By analogy with the hotel example, the price of electricity would be $30 per MWh off-peak and $200 per MWh ($30 + $170) on-peak. Finally, then, assume that during peak periods, the demand for power at the peak price would be 6000 MWh and that the industry has sufficient capacity to meet that demand. Were one to plot what the predicted price of electricity and average variable cost as a function of power demanded, one would then get figure 8.2.

The solid line indicates average variable cost at $30 per MWh; the dots along that line and the solid dot at $200 per MWh and 6000 MWh supplied are the predicted prices.

Already we can see that during peak periods, price will be substantially above average variable cost. However, a number of significant complications must be added to this stylized picture to get a more realistic view of what the competitive supply curve would look like.

1. If a generator is going to be operated only at peak periods, one would expect that it would have a lower fixed-to-variable cost ratio. Since a peak plant will have only a few hours of operation in which it could cover its capital costs, it will be more economical to use low-capital/high-variable-cost technologies at peak periods, with high capital/low variable costs for base-load plants.[14] Thus, one would expect the average variable cost curve to slope upward to some extent as one approaches industry capacity.

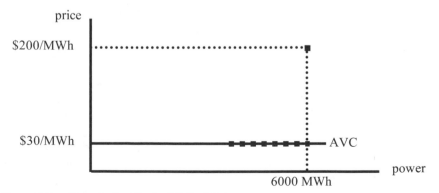

Figure 8.2.   Hypothetical Peak, Off Peak Prices

2. As noted above, the industry's capacity could be exhausted at different levels of demand. This could produce observed price-quantity points filling in the vertical line, at the capacity level of output, between $30 per MWh and $200 per MWh. The extra profits in these "shoulder" demand periods would induce entry, reducing the maximum peak price in this example. However, the possibility of a super-peak demand, reached on fewer than 5 percent of all hours, would tend to increase the maximum price. In any event, the supply curve would tend to have a vertical component as well as a horizontal one, forming a backward "L."

3. There is a separate set of fixed costs in electricity having to do with the costs of starting up and shutting down a generator altogether. A generator may be constructed, but prices will have to exceed not just average variable costs, but produce enough revenue to cover startup and shutdown costs, before a generator will find it appropriate to meet demand.[15]

4. Perhaps most importantly, generators operate in an uncertain environment, in at least two important respects. They do not have perfect knowledge as to what demand will be at a given time. Neither need they know how many of their competitors' generators will be unavailable at full capacity due to scheduled maintenance or unforeseen shutdowns due to equipment failure, transmission congestion, fuel supply, or other contingencies. Thus, one would expect to see generators guess that price may be above variable cost at times when actual supply ends up being below the full industry capacity. This will tend to "fill in" that backward L with observed quantity-price data points. One would expect greater density of data points toward the boundaries of the backward L.[16]

Putting these together would give an observed set of quantity-price data points and an AVC graph that looks something like figure 8.3.

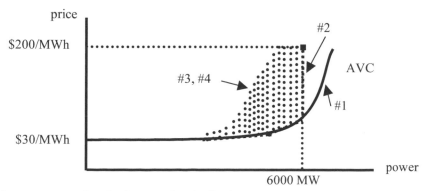

Figure 8.3. Predicted Price, Supply Distribution

The numbers and arrows in the graph indicate the effects listed above. Of course, the dotted area of price and quantity observations does not represent a precise prediction of prices and outputs we would observe. However, the economics of peak-load pricing with a bit of real-world noise could produce patterns like those observed by Short and Swan for Queensland and Victoria.[17]

RESPONSES AND REJOINDERS

One response to these objections to price-cost-based market-power tests is that rates need not be set to recover capacity costs of the last generator. According to this view, generation companies own a portfolio of plants. As FERC has stated, "Amounts earned on the more efficient plants [owned by a generation company] will cover the investment in the marginal plant."[18] But this leaves unanswered the obvious question of why such a company would build and operate that marginal-peaking plant. It would have higher overall profits if it were not to build such a plant, if the generation company would not recover that unit's capital costs.

A second reason one might posit that capital costs should be irrelevant might be that they are recovered in a separate capacity market. If so, one needs at least to describe such a market, including the "strike price" at which the buyer or regulator could demand that excess capacity be brought online. The returns from those sales would then need to be factored in to determine whether the marginal plant is making excessive profits because prices are too high. PAVC-based studies do not incorporate such an analysis.

If capital costs are left out, it should not be surprising that prices will exceed marginal costs, as defined by the average variable cost of the last firm in. But that is consistent with competition. If all firms were identical—of course they are not—price would have to equal minimum average total cost. If capital costs are trivial, then one would not need rents to cover them. Of course, that does not imply that capacity constraints would not be binding during the time it takes to build plants. Scarcity rents do not imply anticompetitive conduct or effect. Again, the question should be not whether prices are higher than they would be if plants could be built instantaneously, but whether suppliers are withholding output to make the prices go even higher.

One possibility, of course, is that the industry is overbuilt, in the sense that the marginal plant is expected to lose money over the long term. However, such an explanation is inconsistent with the theoretical reason for predicting that market power is likely in electricity just because supply and demand are so inelastic. If there is excess capacity because of regulatory or ISO requirements, then the "price" of electricity on- or off-peak would have to include a separate capacity fee component of some kind. These considerations are also typically not present in price-cost-based studies.

THEORETICAL MARKET POWER CONCERNS

Mismeasurement does not imply that generators lack market power, particularly in peak periods. Theory offers some suggestion that regulators might want to be on the lookout for the unilateral exercise of market power, particularly at peak periods.[19] To oversimplify a complicated subject, one can subdivide the possible models into those in which the firms choose prices, and those in which they choose quantities or outputs.

Models based on output may predict noncompetitive outcomes. Such models have two variants, but both are problematic. One variant involves firms undersizing plants to keep output low and prices high. Since capacity choices are made over the long run, this outcome is possible only where fixed costs are large enough and the market small enough to support only a few competitors.[20] These conditions do not appear to hold in the United States.

The second output-based type of model is short-run, taking the number of competitors in the market as given. In this case, generators could make strategic choices to limit production in order to raise prices, taking the supply decisions of the other suppliers as given.[21] With a low elasticity of demand for electricity, this could lead to substantial price-cost margins, even with a relatively large number of competitors. For example, if the elasticity of demand for electricity is -0.2, ten identical generators would, in such a model, set prices equal to double marginal cost.[22]

Such a strategy is not likely to be profitable at peak periods. If we are talking about withholding output below capacity, the relevant marginal cost may well be something like AVC, if marginal costs within the firm are relatively constant. As we have seen, at peak periods one would expect prices to be set at many multiples of AVC. Even excluding the prospect of longer-term entry, firms may be better off producing up to their capacity limits and taking the competitive price, at peak periods, rather than withholding output. Such withholding may be more profitable off-peak, but one would expect the elasticity of demand for electricity to be greater off-peak as well, mitigating the incentive to withhold.[23]

With regard to price-based strategic models, the most important involve "dominant firms" that set prices assuming that competitors will take that price as given.[24] In these models, the price-cost margin at the profit-maximizing price is positively related to the market share of the dominant firm, and inversely related to the elasticity of demand for the product as a whole and the elasticity of supply of its competitors.[25] Off-peak, with capacity exceeding demand, the elasticity of supply of the other firms is likely to be quite high.[26] On-peak, however, the elasticity of demand is likely to be very small, perhaps approaching zero. With a small elasticity of demand as well, a firm with even a small market share may find it profitable to withhold output and set prices substantially above marginal cost.[27]

As with the quantity-based models, on-peak we would expect substantial price-cost margins in any event, so we may not observe such an exercise of market power. We might instead simply see everyone finding it optimal to supply up to capacity and charge prices exceeding marginal (or average variable) cost. However, the possibility of very low supply and demand elasticities at peak periods implies that one cannot dismiss market-power claims, even if a generation market "looks" competitive by conventional measures.[28]

QUANTITY-BASED EMPIRICAL APPROACHES

Some defend price-cost studies at least in part on the basis that critics have not suggested alternative tests for market power.[29] This seems unfortunately to be too much like the "looking where the light is" punch line to the standard economist joke.[30] But there are alternative tests if one does not focus on prices, which under competition can exceed average variable costs, sometimes by orders of magnitude. The focus of assessments of market power in electricity should be not on price but on output, that is, withholding. Specifically, one should seek to identify generation capacity that would have been profitable to run at prevailing market prices but was withheld from sale.

Econometric tools could offer some insight. The increased profits achieved by withholding output accrue not just to the withholder, but to all electricity suppliers. This suggests that all else being equal, a power producer is more likely to withhold capacity the greater its share of overall capacity in the market. That, in turn, suggests a hypothesis: If output is being withheld to exercise market power, one should observe "maintenance shutdowns" disproportionately among producers with larger market shares.

One might see this empirically in two ways. If shutdowns are random, a firm with X percent of capacity should see X percent of shutdowns, all else equal. A simple measure of concentration (e.g., the Herfindahl-Hirschman Index, or HHI)[31] of outages could exceed the measure of concentration of capacity as a whole. More accurately, were one to regress likelihood of outages on firm characteristics, the coefficient of a term relating to market share should significantly exceed one. The larger the firm, the even more likely it is that it would have a generator shutdown.

A potentially better tactic might be to examine output decisions directly, rather than indirectly infer withholding via econometrics. Few industries offer the level of firm-specific cost and output data available regarding the electricity generation sector. If those data are reliable, one should not have to resort to statistics to infer market power. One would not expect generators to be taken off line voluntarily when prices are at their peak. If anticompetitive withholding is going on, the regulator ought to be able to "name names" — identify those generators that have withheld electricity that otherwise would have been profitable to generate if one were taking prices as given. Regulators could investigate specific incidents of peak-period maintenance to see if the output reductions were warranted.

Whether one employs econometric techniques or analyzes specific supply decisions made by electricity suppliers, output data will not be free from ambiguity.[32] Generators frequently need to be taken off line for maintenance purposes and, as noted earlier, the costs of starting up and shutting down units may make generation companies less willing to operate than might seem immediately profitable. Fox-Penner also notes that firms may end up holding capacity in reserve against outages, and such capacity may remain unsold even during a price spike.[33]

To meet an appropriate legal burden before enforcing any policies or punishments, one would need to evaluate other explanations for alleged withholding. Without such careful evaluation, regulators could end up imposing possibly unwarranted mandatory supply requirements on generators.[34] But output-based approaches remain at least as practical and theoretically better alternatives than price-based approaches. A helpful sign is that Joskow and Kahn, among those who have adopted the price-cost method criticized here, are giving more weight to output-based studies.[35]

CONCLUSION

Criticisms of the competitiveness of generation markets are widespread. Unfortunately, many of these criticisms are based on comparisons of prices to average variable cost. However, even in a competitive electricity market, one would expect to see prices substantially above average variable cost during peak demand periods. Variation in demand, increasing average variable cost curves, and particularly uncertainty among generators regarding market demand and supplies from their competitors can give price-cost data patterns not unlike those used to support claims that the industry is not behaving competitively. Last and not least, the prospect of entry could dampen market power over the longer term.

That said, low supply and demand elasticities for electricity, particularly at peak periods, support some degree of concern that generation markets may not be competitive. Better tests for market power would look to quantities, not prices, for example, by seeing if firms with larger market shares are disproportionately more likely to have outages. Perhaps the best test, with the data available, would be for regulators to identify directly the suppliers that seem not to be generating nominally profitable electricity, and then see if any excuses are sound. If regulators attempt to set prices equal to a measure of costs that does not allow firms to earn rents sufficient to cover fixed costs, entry will be discouraged and competition subverted.

Finally, a philosophical observation: The different approaches may arise out of different interpretations of what "market power" is about. From the neoclassical perspective, questions about market power are about looking for efficiency losses, which fundamentally are not based on price but on reduced output. By that criterion, "market power" is fundamentally about withholding.

A less neoclassical perspective may focus on the distributive effect of higher prices. To the extent one cares about distributive effects, one might be drawn more to price than to output. If demand for wholesale power is perfectly inelastic, for example, because retail prices are fixed by regulation (as was the case in California), one could observe higher prices without the output reductions characteristic of the exercise of market power. I would include "price but no output reduction" effects as questions of "market design" or "gaming the auction," but not "market power," a term that I reserve for practices that lead to reductions in supply in order to raise prices and profits.[36] But to the extent one combines those under the same heading, one might be drawn more to price tests for market power. Unfortunately, they do not work very well in addressing that specific concern.

NOTES

1. Essentially, "wholesale" electricity markets are those in which energy is purchased by entities responsible for delivering electricity to those who use it. "Retail" markets are those in which these end users purchase electricity. These definitions are somewhat fluid, in that large industrial electricity consumers may be able and willing to purchase electricity at wholesale, bypassing retail distribution companies. These definitions also are accompanied by legal technicalities, in that whether states or the federal government have regulatory authority over electricity production, delivery, and sale depends on whether those activities are "retail" or "wholesale" in nature. For a more detailed description of this distinction, see chapters 3 and 12 of Timothy Brennan, Karen Palmer, and Salvador Martinez, *Alternating Currents: Electricity Markets*

and Public Policy (Washington, DC: Resources for the Future, 2002). A recent case examining these issues with regard to the U.S. Federal Energy Regulatory Commission's authority over transmission pricing is *New York et al. v. Federal Energy Regulatory Commission et al.*, U.S. Sup. Ct., *slip op.* 00-568, *dec.* March 4, 2002.

2. Richard Green and David Newbery, "Competition in the British Electricity Spot Market," *Journal of Political Economy* 100 (1992): 929–53; John Kwoka, "Transforming Power: Lessons from British Electricity Restructuring," *Regulation* 20 (1997).

3. Paul Joskow, "The California Market Meltdown," *New York Times*, Jan. 13, 2001; Severin Borenstein, James Bushnell, and Frank Wolak, "Diagnosing Market Power in California's Deregulated Wholesale Electricity Market," Working Paper PWP-064, University of California Energy Institute (2000); Paul Joskow and Edward Kahn, "A Quantitative Analysis of Pricing Behavior in California's Wholesale Electricity Market During Summer 2000," Working Paper No. 8157, National Bureau of Economic Research (2001).

4. Christopher Short and Anthony Swan, "Competition in the Australian National Electricity Market," *ABARE Current Issues* (January, 2002).

5. Formally, the Lerner index or price-cost margin is

$$\frac{P - MC}{P},$$

where P is the price and MC is the marginal cost. For a profit-maximizing firm, The Lerner index is typically equal to $1/E$, where E is the elasticity of demand facing the firm. *See* David Kaserman and John Mayo, *Government and Business: The Economics of Antitrust and Regulation* (Fort Worth, TX: Dryden Press, 1995): 101–2.

6. See, e.g., Joskow and Kahn, "A Quantitative Analysis," 5.

7. Also to be clear, the problem is not that marginal or variable cost is assumed to be increasing within the capacity range of the generator itself. For simplifying purposes, we can assume that average variable cost (hence marginal cost) is constant within a particular generator, until it hits its capacity limit.

8. Short and Swan, "Competition," 2. Short and Swan's discussion here is somewhat ambiguous, because just before these sentences, they state that "where marginal costs can increase quickly as demand approaches capacity limits, competitive prices can exceed the marginal cost of producing the required electrical energy." However, they go on to state that prices must be less than "marginal cost of an additional unit of energy from another generator." This is not correct, as we see below.

9. This point is not new. See Robert Dansby, "Capacity Constrained Peak Load Pricing," *Quarterly Journal of Economics* 92 (1978): 387–98, especially 394.

10. The actual level of the on-peak price in the short run would be above or below this value, depending on whether demand was higher or lower than that expected when the unit was constructed. Leaving aside earning premiums to cover risk, if demand turns out to be greater than anticipated, the firms would earn long-run rents as the market-clearing price exceeds the expected level at which revenues would cover costs. If demand is lower than anticipated, the reverse holds—the market-clearing price is lower than that which would have covered cost.

11. Some natural gas power plants in Southern California had to purchase permits to emit nitrous oxides, a pollutant that contributes to the formation of particulates and ground-level ozone. Because the supply of such permits was fixed to limit nitrous oxide emissions, as electricity prices in California rose, so too did permit prices, until regulators stepped in and substituted a fixed price for permits for the supply limitation. Joskow and Kahn ("A Quantitative Analysis,") treat the permit price as an exogenous price that should be included in calculating the average variable cost of the marginal gas plant in estimating its price-cost margin. However, that permit price is not exogenous but endogenous. Even if electricity were being withheld, one would think that the price of nitrous oxide permits would be driven up to the difference between the market price of electricity and the marginal cost of a gas plant, taking into account the marginal emissions of that plant. That Joskow and Kahn find a difference between price and "marginal cost" (actually average variable cost of the marginal plant) even including these permit prices suggests that there is a difference between actual and measured marginal cost that their empirical procedures neglect.

12. Federal Energy Regulatory Commission, Order on Rehearing of Monitoring and Mitigation Plan for the California Wholesale Electric Markets, Establishing West-Wide Mitigation, and Establishing Settlement Conference, EL00-95-031 et al., issued June 19, 2001, 27–28, 34 (hereafter referred to as "FERC Mitigation Order").

13. Imagine that under competition N firms would enter, and rank firms from lowest to highest operating cost from 1 to N. Under the FERC Mitigation Order's pricing rule, the Nth firm, with the highest operating cost, would not be allowed to ever set prices that would enable it to recover its fixed costs. Thus, it would not enter, making the $N-1$ firm the highest-cost firm. But the rule would now reduce the maximum allowed price, so that the $N-1$ firm is now unable to cover its fixed costs, and thus it would not enter. With rational expectations by entrants, the situation would unravel to the point that no one of the N firms would enter.

14. Michael Crew and Paul Kleindorfer, *The Economics of Public Utility Pricing* (Cambridge, MA: MIT Press, 1986).

15. When these, along with ancillary service revenues, are factored into profitability estimates, the decision as to whether a plant is profitable to operate at a seemingly high price becomes extraordinarily problematic. See Scott Harvey and William Hogan, "Identifying the Exercise of Market Power in California," Dec. 28, 2001, available at www.ksg.harvard.edu/hepg/Papers/Hogan%20Harvey%20CA%20Market%20Power%2012-28-01.pdf, especially 11–26.

16. One error avoided in some price-based studies of the California situation is that prices would be inflated by the prospect that bankrupt distribution utilities would not honor their promises to pay for wholesale energy, particularly when the courts mandated such sales to prevent blackouts. On this basis, James Bushnell does rely on California price data only during the winter of 2000–2001, when prices were at their highest despite off-peak demand, to support market power claims. James Bushnell, "What We Do and Do Not Know About How Electricity Markets Work," keynote address, NEMS/Annual Energy Outlook 2002 Conference, Crystal City, VA (Mar. 12, 2002), notes available at www.eia.doe.gov/oiaf/aeo/conf/pdf/bushnell.pdf.

17. Short and Swan, "Competition," 5–6. Their specific test for whether a market was not competitive if the median Lerner index over a given month exceeded .3. They

are right to suggest that the pattern for Victoria is more likely to be suspicious than that for Queensland, but not necessarily because Victoria has more observations a certain percentage above average variable cost of the marginal plant. Rather, it is because the observations for Victoria appear to have more points farther above average variable cost *and to the left of the vertical "capacity" or "maximum output" line*. Such observations require greater amounts of seemingly profitable electricity to remain unsupplied. It is the quantity that matters, not the price. We return to this point below.

18. FERC Mitigation Order, 34.

19. For a related discussion of the theory and other issues raised regarding market power in electricity, see Timothy Brennan, *The California Electricity Experience, 2000–2001: Education or Diversion* (Washington, DC: Resources for the Future, 2001): 37–40.

20. In some industries, firms may not enter even when prices are high because they would predict that the added competition resulting from their entry might result in prices so low that they would not recover sunk costs incurred by coming into the market. This is especially true for industries in which fixed costs are so great that competition among just a few firms would generate too little revenue to cover them. Hence, one cannot say as a matter of absolute generality that entry cures all ills, and the prices can never be sustainable above competitive levels. However, the opening of electricity generating markets around the world has been predicated on the view that fixed costs and resulting economies of scale are not so large as to result in only a very small number of suppliers being viable.

21. Limiting or withholding output could be accomplished either by simply not producing energy, or by offering to sell it only at prices above what buyers are willing to pay. On either account, the result, a reduction in supplies actually purchased in order to drive up prices, is the same. We may also have situations in which prices are bid up, taking advantage of the design of the electricity market, but with no power being withheld. Pure price-manipulation situations should be analyzed as failures of the design of the market, and not as the anticompetitive exercise of market power. Brennan, *The California Electricity Experience*, 33–35.

22. The standard result for identical firms in a quantity-based model is that the Lerner index or price-cost margin (see note 5 above) equals $1/NE$, where N is the number of firms and E is the absolute value of the elasticity of demand.

23. Obviously, this is not necessarily true. However, at any price, off-peak quantity demanded will be lower than on-peak quantity demanded at any price. If the slope of the demand curve at that price is the same off-peak as it is on-peak, then the elasticity off-peak will be greater. Of course, the off-peak demand curve could be steeper than it is on peak, producing an outcome counter to this intuition.

24. William Landes and Richard Posner, "Market Power in Antitrust Cases," *Harvard Law Review* 94 (1981): 937–96.

25. Formally, the expression is

$$\frac{P - MC}{P} = \frac{S}{E_D + [1 - S]E_S},$$

where S is the firm's market share, E_D is the absolute value of the elasticity of demand, and E_S is the collective supply elasticity of the other firms.

26. Because electricity is by and large a homogeneous good, when firms are choosing prices, the noncooperative (Bertrand) outcome is likely to be competitive, e.g., price equal to marginal cost, when capacity constraints do not intervene. See Kaserman and Mayo, *Government and Business*, 187–88; one can get higher prices when goods are differentiated. See Eric Rasmusen, *Games and Information* (Cambridge, MA: Blackwell, 1994): 314–18. In the market, electricity could be a differentiated product, e.g., if some consumers are willing to pay a premium for so-called "green" power. See Timothy Brennan, "Green Preferences as Regulatory Policy Instrument," *Ecological Economics*, vol. 56, no. 1 (2006): 144–54..

27. For example, applying the equilibrium condition in note 25 above, we might observe a firm with 10 percent of the market (S = .1) setting price at about 50 percent (10/19) above its marginal cost if the elasticity of demand is .2 and the elasticity of supply of the other firms is .1.

28. The prospect of profits exceeding competitive levels would be expected to attract entry. But under the very low supply and demand elasticities associated with electricity at peak periods, it is not clear whether such entry would preclude the exercise of market power, even in the long run, or whether it would merely spread the profits among a larger set of firms. If the latter is the case, entry might depress profits per firm, but not peak-period prices.

29. Severin Borenstein, "The Trouble with Electricity Markets: Understanding California's Restructuring Disaster," *Journal of Economic Perspectives* 16 (2002): 191–212.

30. For those who may not know the joke, it begins with an economist crouched under a streetlight at night, looking for something. A second person walks over and asks, "Do you need any help?" The first responds, "I'm looking for my car keys."

"Where did you leave them?"

"Over there," the economist replies, pointing down the street.

"Why are you looking here, then?"

"Because this is where the light is."

31. The HHI is the sum of the squares of the market shares of sellers in the market. In a monopoly market, where one firm has 100 percent, the HHI is 10,000. In an atomistic market, as market shares approach zero, so will the HHI. If there are N equal-sized firms in a market, the HHI will be 10,000/N. For a discussion of how the HHI is used in electricity mergers in the United States, see Federal Energy Regulatory Commission, "Inquiry Concerning the Commission's Merger Policy under the Federal Power Act: Policy Statement, Order No. 592 "Docket No. RM96-6-000 (December 18, 1996), especially Appendix A, 59–62, 74–79, available at www.ferc.fed.us/Electric/mergers/mrgrpag.htm#PolicyStatement.

The HHI has three theoretical rationales, none especially compelling. A first, due to Stigler, is a model in which the likelihood that a firm in a cartel would detect that a loss of market share is the result of nonrandom price cutting by someone else rather than random variation in where customers shop is a function of the HHI. George Stigler, "A Theory of Oligopoly," in George Stigler, *The Organization of Industry* (Homewood, IL: Irwin, 1968). A second is that in a Cournot model, the HHI is proportional to the welfare increase attainable from increasing industry output by a fixed amount. Robert Dansby and Robert Willig, "Industry Performance Gradient Indexes," *American Economic Review* 69 (1979): 249–60. Third, in a Cournot model with constant demand elasticity in which firms have different marginal costs, the mark-up of price over a share-weighted average of marginal cost will be proportional to the HHI.

32. For detail, see Harvey and Hogan, "Identifying," note 15 above, at 47–71.

33. Peter Fox-Penner, Comments Before the Federal Energy Regulatory Commission, Investigation of Terms and Conditions of Public Utility Market-Based Rate Authorization, Docket No. EL01-118-000 (Feb. 11, 2002): 44.

34. See FERC Mitigation Order, 8, 12–18, ordering that generators in the western United States "must offer" unscheduled non-hydroelectric capacity unless committed to maintain minimum operating reserves.

35. Paul Joskow and Edward Kahn, "A Quantitative Analysis of Pricing Behavior in California's Wholesale Electricity Market During Summer 2000: The Final Word," mimeo, Feb. 4, 2002.

36. See note 21 above.

9

The Role of Distributed Energy Resources in a Restructured Power Industry

David E. Dismukes

In 1978, the federal government passed the Public Utilities Regulatory Policies Act (PURPA) and dramatically altered the nature of the electric power industry. PURPA refuted the notion that electricity generation was a natural monopoly and opened the door for future competition. During the same period, advances in generation technologies altered the existing planning paradigm that large central-station electric generation facilities, like nuclear and coal, were the most effective and least costly way of meeting electricity demand. Small-scale, modular, and highly efficient combustion turbines and combined-cycle units came to compete with, and in many cases replace, large generation units.

Even after more than twenty-five years, the electric power industry still finds itself in the position of redefining its organization, players, and set of regulations. This is no more apparent than in the considerable industry restructuring, and re-restructuring, that has occurred since 1996. Over the past decade, the industry has gone through a variety of much-heralded "paradigm shifts" only to find that in many ways, the more things change, the more they stay the same.

DISTRIBUTED ENERGY RESOURCES

One of the great paradigm shifts heralded over the past decade's experience with restructuring is associated with distributed energy resources, also know as "DER," or "DG" ("distributed generation") for short. DER refers to generation, storage, and demand-side management (DSM) devices and technologies that are connected to the electric grid at the distribution level (i.e., below the bulk power transmission system). These devices and technologies include microturbines, reciprocating engines, fuel cells, wind turbines, photovoltaics, and flywheels. Because these devices are more modular and flexible than a large central power station, they can be located at the customer's premises on either the utility side or the customer side of the meter.

One of the significant benefits claimed by DER advocates is that it provides the electricity consumer with greater reliability, higher power quality, and more flexible choices. Smaller users of electricity have started to realize what large volume customers have known for some time: there are important qualitative aspects to electricity service. Traditionally accepted levels of power service quality have been fine for many applications over the past half century, particularly those that are electro-mechanical in nature. But today's computer server-farms, and a large portion of high-tech manufacturing, require much greater reliability and quality than traditionally accepted (and provided) levels of service. In addition to service benefits, many DER technologies can be more environmentally friendly and less expensive than the provision of typical utility service. Widespread use of DER technologies could also mitigate congestion in transmission lines, help to control price fluctuations, and provide greater stability to the electricity grid.

Despite the promise of DER, it has failed to take off as a competitive opportunity for various types of end users. The reasons for these failures are in many ways consistent with the transitional problems being experienced by a wide range of competitors (or former competitors) in U.S. energy markets. The problems contributing to the near-term downturn of DER include:

- A glut of high-efficiency combined cycle generation that, in many regions of the United States, is sitting either idle or running at less than optimal levels due to a variety of factors.
- A recession, and subsequent lackluster recovery for high energy-use sectors, particularly the high-tech sector targeted by many DER providers due to its need for high reliability power.
- Failure of the regulatory process to continue to press for market reforms and the development of truly widespread independent organizations of market governance.

While these problems have created serious development issues for DER, they are transitional in nature. Changes in the economy, extreme weather conditions, and a renewed regulatory commitment to competitive market development could turn many of these factors around in relatively quick order.

What is more problematic are the ongoing structural issues associated with DER: namely, the challenge it will continue to provide to incumbent utilities, even those subjected to ever-increasing degrees of competition. Paradoxically, although DER technologies are a way to become less dependent on the grid, they will not be widely adopted until they can also be connected to the grid. Users of DER will want access to the grid not only to buy electricity but also to sell it. Therefore, maximizing the benefits of DER will require careful adjustments in the current regulatory environment.

The Potential Benefits of DER

It is useful to begin with a "vision" that many advocates have for DER in the future. Over the past thirty years, large sectors of the U.S. and global economies have moved from mass production to flexible or customized production. Automakers like Toyota now run just-in-time assembly lines that produce not simply "a car" but exactly the car that was ordered by an individual customer, and is to be delivered to that customer within a two-week period. While mass U.S. steelmakers are rusting away, small-scale firms that produce specialized steel are flourishing. The Internet and new printing technologies allow books to be produced on demand, resulting in greater choice for customers and more room for promising new authors. DER technologies can be understood as creating analogous changes in the electric power industry.

In an extreme vision, DER technologies would replace current mass production methods altogether. The attraction of this vision is that when generation, transmission, and distribution are all local, the peculiar problems of the electricity industry created by economies of scale, difficulties of storage, and grid externalities disappear. Power generation in the DER setting becomes a competitive industry much like any other. The extreme scenario is of course unrealistic, but it does point toward some of the advantages of a DER-accommodated system.

In particular, DER has the potential to reduce demands for transmission and distribution infrastructure, the two areas where problems associated with monopoly power, allocating fixed costs, pricing, independent governance, and long-run investment stymie efforts to efficiently restructure. Additionally, transmission infrastructure development is often stunted by the not-in-my-backyard (NIMBY) attitudes of local interest groups. To the extent that DER can reduce transmission infrastructure demands, it may also reduce the need for difficult transmission planning decisions especially in dense urban areas facing considerable costs for new, expanded transmission expansions, not to mention the NIMBY concerns associated with infrastructure development.

These broad examples highlight the impact DER can have on the status quo electric utility industry. Some incumbents, recognizing the threat of DER, have tried to "co-opt" the idea by arguing that DER is best developed by incumbent utility distribution companies (UDCs or "utilities") (Lesser and Feinstein, 2002). This approach, however, could lead to the strangling of DER in its infancy, since it perpetuates the ongoing problem associated with incumbent utility ownership of generation and transmission assets and the inherent incentives to give preference to one, to the detriment of the other.

DER applications also stimulate entrepreneurship and product differentiation. If DER technologies can approach the costs of mass production, then the demand for power will likely differentiate greatly in the dimensions of reliability, quality, and environmental standards. Providing power services that customize all of these needs facilitates the customer choice initiative being pursued throughout the United States.

DER technologies, however, are not just a substitute for the grid. If efficiently developed, these technologies can lead to a more robust and diversified electric power grid. These technologies can also complement many commercial aspects of a restructured power market. In many instances, DER technologies can be brought on- and off-line at a lower cost than traditional technologies, which makes DER power production more sensitive to price signals. Increasing the elasticity of the power supply makes for a more robust system and one that is less capable of being gamed by marginal producers (Borenstein 2002).

The DER Technologies Available

While DER refers to a broad range of technologies and applications, most attention is being directed at those opportunities to generate on-site electricity. Among the technologies available to suit small- and medium-sized DER applications, those fueled by fossil energy include conventional steam turbines, combustion turbines, reciprocating/internal combustion engines, microturbines, and fuel cells. Renewable technologies include photovoltaic cells, wind-powered generators, and biomass-fueled generators. Table 9.1 presents the cost and operating characteristics of these DER technologies.

Conventional steam turbines and combustion turbines are mature technologies generally used for medium- and larger-sized applications. As a preferred technology for most conventional generation applications requiring more than several megawatts of power, the popularity of these turbines is based upon their quick development times, proven reliability and availability, low emissions, quick environmental permitting times, and low maintenance and operating costs in comparison to other DER applications. Small combustion turbines are found in a wide variety of applications including mechanical drives, base-load grid-connected power generation, peaking power, and remote off-grid applications. The installed capital costs of these technologies range from $300 per kilowatt (kW) to $1000 per kW and tend to increase with

Table 9.1. Examples of Selected DER Costs and Characteristics

	Capacity (kW)	Capital Cost ($/kW)	Fuel Cost ($/kWh)	O&M Cost ($/kWh)	Service Life (Years)	Heat Rate (Btu/kWh)
Microturbine—Power Only	100	1,485	0.075	0.015	12.5	13,127
Microturbine—CHP	100	1,765	0.035	0.015	12.5	6,166
Reciprocating/ICE—Power Only	100	1,030	0.067	0.018	12.5	11,780
Reciprocating/ICE—CHP	100	1,491	0.027	0.018	12.5	4,717
Fuel Cell—CHP	200	3,674	0.029	0.010	12.5	5,106
Solar Photovoltaic	100	6,675	—	0.005	20.0	n.a.
Small Wind Turbine	10	3,886	—	0.005	20.0	n.a.
Large Wind Turbine	1,000	1,500	—	0.005	20.0	n.a.
Combustion Turbine—Power Only	10,000	715	0.067	0.006	20.0	11,765
Combustion Turbine—CHP	10,000	921	0.032	0.006	20.0	5,562
Combined-cycle System	100,000	690	0.032	0.006	20.0	5,642

Source: Congressional Budget Office, *Prospects for Distributed Electricity Generation*. September 2003.

decreasing power output. Compared with reciprocating engines, combustion turbines tend to cost more for smaller sizes and less for larger sizes.

Reciprocating engines are the most mature and commonly used technology for distributed generation. They are generally less expensive than competing technologies and have start-up times as low as ten seconds, compared to emerging technologies that may take hours to reach steady-state operation. The capital cost of a basic gas-fueled generator set package ranges from $300 to $900 per kW, depending on size, fuel type, and engine type. Additional costs include balance of plant equipment, installation fees, engineering fees, and other owner costs. As the traditional technology for emergency power, groups that have been interested in these technologies to date include small-scale industrial and manufacturing customers, oil and natural gas producers, hospitals, telecommunications switching stations, and Internet server farms.

Microturbines and fuel cells are two technologies that are attempting to challenge the reciprocating engine market. Microturbines are essentially small jet-aircraft engines that are in many ways similar to the technologies that revolutionized the larger combustion turbine market of the electric power industry over the past ten years. Fuel cells, on the other hand, facilitate a chemical process that makes electricity rather than storing it. Unlike many energy conversion processes, fuel cells generate electricity in a continuous direct process, thereby reducing the excessive energy losses prevalent in other multistep energy conversion applications.

Both microturbines and fuel cells offer a significant number of advantages over conventional reciprocating engines that make them very attractive for small-scale applications. Microturbines can be used for stand-by power, power quality and reliability, peak shaving and cogeneration applications. Microturbines produce between 25 and 500 kW of power and are well suited for small commercial building establishments such as restaurants, hotels, small offices, and retail stores.

Fuel cells are unique in terms of the variety of their potential applications. Fuel cells have several benefits over other DER applications as they operate quietly and reliably, are extremely efficient, and produce much smaller amounts of greenhouse gases. If pure hydrogen is used as a fuel, fuel cells emit only heat and water as a by-product. One of the driving forces behind fuel cell development is the emphasis on moving toward a hydrogen-based economy. The United States government and the fuel cell industry have invested hundreds of millions of dollars in developing technology that will compete with conventional power production.

One of the typical end-user benefits associated with DER applications rests with these various technologies' ability to let customers choose their own level of reliability and cost. Today a number of retail customers are becoming increasingly more sophisticated in their power needs and requirements. Some of these customers could benefit by moving away from "plain-vanilla" service as has been traditionally provided by utilities. While examining these trade-offs, it is important to consider that in the past, the degree of "economic" and "reliable" power has been determined predominantly by a regulatory process. (See, for example, Woo, Horowitz, and Martin 1998.) Even today, as more retail markets become increasingly competitive, historic traditions in service provisions remain constant, particularly when it comes to customers taking service under default, or "provider of last resort" terms.

Figure 9.1 presents the reliability/cost trade-offs in service provision between traditional service[1] and DER applications. Both curves show that for a low-cost level, extremely economical power can be provided at the expense of reliability. As the curves move upward, more reliable power can be provided, but the additional reliability increases the cost of electricity production. In the middle region of both curves, there are various degrees of cost and reliability. Since traditional service usually draws electricity from either utility-owned central-station generation or competitive wholesale markets, there are some benefits to this type of service when the end user demands a moderate trade-off between cost and reliability. However, at the outer extremes, the DER application can outperform traditional service in either category.

Consider a very low-cost application that does not provide very reliable power. For some types of end users, rebuilt or used generating equipment

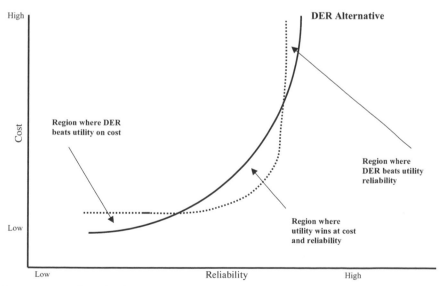

Source: H. Lee Willis and Walter G. Scott. (2000) *Distributed Power Generation: Planning and Generation*

Figure 9.1.   Cost-Reliability Trade-Off Between DER and Utility-Provided Power

may suffice to generate low-cost electricity at nonessential facilities. In this instance, when cost, and not reliability, is everything to the end-user, DER can serve a very useful purpose.

At the upper ends of the curves, DER can provide more reliable power than the grid for end users, but this reliability will come at a cost. For end users with critical applications, such as Internet and computer servers, telecommunication switches, and hospitals, the economics of higher-cost applications make sense given the cost that reliability failures have on operations. Figure 9.1 shows where many readily available DER markets are today. Over time, as DER applications become more cost effective at moderate and higher levels of reliability, both curves will tend to fall upon one another.

In addition to the reliability-cost trade-off discussed above, there are a number of other types of applications and trade-offs that can be considered different for DER providers and customers. Table 9.2 presents the different applications and stakeholder groups that can be impacted by DER—and the advantages of these applications to the various groups.

Despite these numerous opportunities, there are, and continue to be, a number of barriers to DER applications—many of which are similar to the barriers faced by industrial cogeneration applications allowed under PURPA in the early 1980s. Several barriers are the result of traditional regulation, and are being impacted by the retail restructuring process unfolding in a number of different states.

Disincentives Associated with Developing DER Applications

The impact that regulatory policy has upon DER cannot be emphasized enough. DER resources create the largest opportunities if they are interconnected to the distribution grid. The benefits of DER deteriorate rapidly for nongrid applications, but attempting to maximize the opportunities for cost-effective DER is difficult since most UDCs are faced with a number of serious disincentives for supporting interconnection.

These disincentives, in large part, are a function of distribution-level regulation. Specifically, the loss of sales on every kWh of self-generated electricity is a kWh of sales that has been lost by the utility or its affiliate. These revenue losses can translate directly to lost profit opportunities. Additionally, the potential precedent-setting nature of DER creates strong incentives for UDCs to prevent any of the applications from ever getting off the ground. Clearly, an individual DER project of less than 1 MW can hardly constitute a major earnings threat to a single utility. However, the cumulative impact of numerous projects concentrated in a particular utility service territory could have

Table 9.2. Impact of DER on Various Stakeholders

Stakeholder Group	Combined Heat & Power	Standby Power	Peak-Shaving	Grid Support	Stand Alone
Customer	Lower energy costs, higher overall reliability.	Avoid economic loss due to system outage and satisfy critical support systems.	Lower peak-period energy costs.	Customers generally benefit from the enhanced service provided but may be isolated from competition markets as a result.	Customer option to avoid high-cost backup service, remote communications, and control systems.
T&D System	Positive to negative depending upon the application.	Can be integrated with utility needs to provide both customer and grid benefits.	Can be integrated with utility needs to provide both customer and grid benefits.	Enhances grid stability and economic customer service.	Loss of customer load and associated revenues.
Energy Service Provider	Power and heat can be separately marketed; ESPs can also provide ancillary services to CHP customers.	Can facilitate ESP marketing of interruptible power supplies; widely used strategy of municipal systems.	Can aggregate and sell customer peak-period generation.	Possible benefits as an owner/operator of the system.	Possible benefits as an owner and operator of the system.
Natural Gas Industry	Benefit from high gas consumption, possible fuel switching benefit for oil-fired boilers.	Minimal impact, but cost to service customers is high.	Good match of gas off-peak period with electric on-peak period.	Generally similar to peak-shaving benefits.	Benefit from high gas consumption.
Society	Environmental benefits with some technologies, energy efficiency, economic development.	Public health and safety.	Environmental and energy efficiency benefits.	Environmental and energy efficiency benefits.	Less likely in a competitive market to represent an optimum allocation of resources.

Source: Gas Research Institute, "The Role of Distributed Generation in Competitive Energy Markets," Distributed Generation Forum (Chicago: Gas Research Institute, 1999), 9.

significant impacts on growth opportunities. Therefore, many UDCs will attempt to prevent projects from occurring due to the potential precedent-setting nature of the single project.

Some of the most common barriers for DER are related to physical interconnection terms for DER applications, rate design, distribution-level wheeling, and stranded costs. The following sections consider each of these fundamental disincentives and their implications for DER implementation.

Interconnection Issues

One of the major barriers associated with DER implementation has been the inability to easily and seamlessly interconnect with the utility distribution grid. This interconnection can occur at either the transmission or distribution level. Typically, larger applications have a high probability of being tied to the grid at the transmission level (115 kilovolts and above). For most small-scale applications, interconnecting a distributed resource at the distribution level would entail hooking the application to a three-phase distribution line at the 69-kilovolt level.

DER interconnection is normally controlled by UDCs since most applications are interconnected at the distribution level. Unlike large industrial cogenerators, there are no federally mandated requirements for interconnection, emergency and standby power, and buyback rates for excess on-site power.[2] In states where regulators have not required utilities to develop favorable DER rules, the challenge is even more difficult since interconnection, at this point, is dictated by the host UDC.[3]

As noted earlier, it is important to keep in mind that interconnected DER applications offer the electricity grid the greatest benefits. If these applications are discouraged through the interconnection process, there will be a socially and economically inefficient level of DER developed. Some of the more significant DER interconnection barriers include unreasonable interconnection terms for projects, delayed interconnection studies and high fees, and expensive facilities upgrade costs.

Interconnection requirements for DER can often be as complicated as a large merchant generation facility. It seems unreasonable for a 200 kW DER project to go through the same administrative hurdles for interconnection as a 200 MW merchant facility interconnected at the transmission-system level. Equally important are timing issues associated with understanding the interconnection requirements, processing the interconnection request, and beginning the initial study process. Some states, in particular Texas and California, have tried to reduce these barriers through a number of streamlined rules to reduce informational and administrative costs for DER applications. California established "Modified Rule 21" that was designed to streamline the interconnection of new, small-scale generating facilities thereby relieving California's electricity supply constraints and encouraging self-generation while maintaining a comprehensive and user-friendly application form.

Under California Rule 21, the utility does not approve or disapprove of a design, but rather determines if the design complies with requirements. The interconnection rules are based on a screening process that determines the level of review process for interconnected systems. After DER operators apply for interconnection, the utility performs the "Initial Review Process" of the project plans. If all screens are passed, then the system qualifies for "Simplified Interconnection," whereby no additional studies are needed. If a system does not pass the initial review, it must undergo a "Supplemental Review Process." As an outcome of the supplemental review, systems may be permitted to undergo "Simplified Interconnection" with some additional requirements, or where one or more screens are not passed, the system must undergo a formal "Interconnection Study," for which the costs are determined by the utility and borne by the system owner. If the generator accepts the results of the utility's review, then an interconnection agreement is executed and the applicant installs the generator and interconnect equipment.

In addition to complying with Rule 21, interconnections must also comply with local, state, and national codes such as the National Electric Code and the Uniform Building Code. Technical provisions for DER installations include requirements regarding voltage and frequency fluctuation, flicker, harmonics, islanding, DC injection, and protection devices.

According to the Database of State Incentives for Renewable Energy ("DSIRE," funded by the U.S. Department of Energy), forty states have interconnection standards in place. While some of these states' regulatory commissions have not actually installed formal rules, the states' utilities have developed DER interconnection agreements. The following is a short summary of a few selected states and their DER interconnection arrangements at the time of this writing:[4]

- Arkansas: On July 26, 2002, the Arkansas Public Service Commission (PSC) approved final Net Metering Rules. Section 3 of the Rules applies to the interconnection of net-metering facilities to existing electric power systems. Facilities producing electricity using solar, wind, hydro, geothermal, and biomass resources are eligible to interconnect and net meter. Microturbines and fuel cells using renewable resources are also eligible. Customers must submit a standard interconnection agreement to the utility thirty days prior to interconnecting. The facility must meet all performance standards established by local and national electric codes, including the National Electric Code, the Institute of Electrical and Electronic Engineers, the National Electrical Safety Code and Underwriters Laboratories. In addition, utilities may require facilities to meet any other safety and performance standards approved by the PSC.

- Arizona: Statewide interconnection rules have not been established in Arizona, although the state's utilities have individually developed DG interconnection agreements. In 1998–1999, the Arizona Corporation Commission (ACC) convened a DG working group to establish recommendations on interconnection. The draft rules were released in November 1999, and the working group report was released the following summer. The ACC, however, has not formally acted on this.

- Florida: Florida PSC Administrative Rule 25-6.065 allows interconnection of small photovoltaic systems up to 10 kW. The rule applies to investor-owned utilities in Florida, but not to municipal utilities or rural electric cooperatives. It does not specify any limit on enrollment for each utility. Utilities such as Gainesville Regional Utilities (GRU), Gulf Power, and Florida Power & Light (FPL) have filed standard interconnection agreements with the Florida PSC. These agreements require interconnected customers to comply with Underwriters Laboratories (UL) and the Institute of Electrical and Electronic Engineers (IEEE) safety standards for PV modules and inverters. All customers must have at least $100,000 in liability insurance for interconnected systems. Rule 25-6.065 allows each utility to specify within its standard interconnection agreement whether an external manual disconnect switch is required. Both Gulf Power and FPL require customers to install this equipment at the customer's expense.

- Illinois: There are no statewide interconnection rules in place for DER, but the Illinois Commerce Commission has studied the interconnection issue internally. No formal policies or orders have been developed; however, the difficulties encountered by distributed generation developers or owners as a result of a lack of simplified statewide rules have been documented. Individual utility rules are in place to address the interconnection of small-scale DG. In the case of the state's largest utility, ComEd, DG rules are outlined in its *The DG Book: Guidelines for Interconnection of Distributed Generation to the ComEd System*.

  ComEd's rules segment DG equipment into the following capacity categories: 25–2500 kVA, 2500–10,000 kVA, and more than 10 MVA. In general, the system owner is responsible for all interconnection study charges, and systems must be a PURPA-qualifying facility to receive any payment for power sent to utility, with the exception of net-metered PV and wind systems under 40 kW. The difference in interconnection requirements for systems in the capacity categories is in the specific relay standards. Procedurally, all systems require a series of reviews with

ComEd engineering staff. It is worth noting however, that ComEd does not allow interconnection of DG within the "Loop," which comprises the heart of downtown Chicago. As a result, a number of potential DG systems as backup power in downtown high-rises have not been able to interconnect.

- New York: New York became the second state to issue uniform interconnection standards for DER when the PSC originally adopted the Standard Interconnection Requirements (SIR) for units of 300 kVA or less. These rules were issued in December 1999. Because of concerns over some of the burdensome procedural issues, the PSC revisited the rules and, on November 6, 2002, issued an order adopting several modifications to the SIR. The changes streamlined the application process and provided a more ordered progression for the study and review phases of the procedure. As a result of the order, the design review phase was eliminated, and the overall processing time was reduced by ten to twenty business days.

  The SIR addresses technical guidelines for interconnection and application procedures, although it leaves many details to the discretion of utilities. It includes simplified requirements for small systems that qualify for net metering. (Prior to the November 2002 order, interconnection standards for net-metered systems were separate from the DG standards in the SIR.) Procedurally, the standard includes an eleven-step process that covers from initial inquiry to final utility acceptance for interconnection. Included in the appendixes of the SIR are a standard contract and standard application forms.

- Texas: Texas completed its standardized DER interconnection requirements through two different rule makings. The first interconnection rule explains the technical requirements for interconnection and parallel operation of on-site distributed generation. In general, the rule lists requirements and procedures necessary for safe and effective connection and operation of distributed generation.

  The second Texas DER interconnection rule identifies the terms and conditions that govern the interconnection and parallel operation of on-site distributed generation. The rule notes its enabling legislation, the Public Utilities Regulatory Act of Texas (PURA), entitles all Texas electric customers access to on-site distributed generation. It also calls for the establishment of technical requirements to promote the safe and reliable parallel operation of on-site distributed generation resources, and for the promotion and use of distributed resources in order to provide electric system benefits during periods of capacity constraints.

Rate Design Issues

While interconnection is both a physically and economically important issue, rate design can also have significant implications for DER applications. Rate design may seem to be an anachronism in a restructured market; however, most retail restructuring plans will continue the traditional regulatory structure for distribution service. These rates will be important in sending signals to potential DER projects. If these rates are developed in a relatively inefficient manner, a socially suboptimal level of DER will be developed. Most retail restructuring plans, while opening the generation portion of the market to market-based rates, will continue to attempt to balance efficiency and equity issues at the distribution level. The following subsection will discuss how optimal distribution level tariffs would impact DER development, while the later sections discuss the realities of the rate-making process.

THE DEVELOPMENT OF AN OPTIMAL DISTRIBUTION LEVEL TARIFF

An optimal rate design would consider a distribution delivery rate, or family of rates, that varies by:

- time (i.e., real-time pricing or some type of peak/off-peak price differentiation);
- location (load relevant to local capacity, growth, and any congestion constraints);
- desired firmness of capacity (the allowable rate of service interruptions); and
- volume (the amount of energy consumed).

These options would allow all customers of the distribution system to select the type of service each desires and influence their own cost of service. When the time dimension, reflecting system loading, is finely differentiated (for instance, by real-time pricing), the need for standby, maintenance, emergency, or curtailable tariffs is less significant. Customers will pay for the delivery service when it is used, where it is used, with respect to firmness of delivery, and how much is needed. As we move away from a finely differentiated real-time-loading dimension into one measured on an hourly or quarter-hourly basis(peak/off-peak operation), we begin to require the introduction of other tariffs, such as standby and interruptible service, to account for the ways that customers might "group" themselves together.

Ultimately, unbundled distribution service customers should be free to select the type and amount of distribution service that matches their energy needs. A one-size-fits-all approach to distribution service is inefficient for DER and all energy efficiency applications. This approach however, can serve as an impediment to customer choice since DER customers rely less heavily on the distribution system than full-requirements customers. One would expect that in a perfectly competitive world, DER customers would pay less for using the distribution system. Failure to recognize these differences in usage can result in fewer DER applications being developed, and as a consequence, limiting an important competitive energy service choice for end users.

While finely differentiated rates may make considerable sense from an efficiency perspective, it must be recognized that the ongoing rate-design practices of most regulatory commissions pursue a number of other important policy considerations that prevent a movement to perfectly differentiated pricing structure from occurring. These include the goals of keeping rates equitable, affordable, and understandable for all customers.

In addition to public policy goals, there is also the equally difficult challenge of developing complicated and finely differentiated distribution-level tariffs that would be both understandable and administratively cost effective. Because of these "realities" of regulatory rate making, it seems likely that most DER applications will have to settle for discrete, yet differentiated, distribution rates that do take the nature and costs associated with the various types of distribution service into consideration.

SHIFTS FROM VARIABLE TO FIXED CHARGES IN DISTRIBUTION TARIFFS

In the past, traditional utility tariffs were developed in a manner that reflected a two-part tariff composed of a fixed customer charge and a per kWh energy charge. The usage-sensitive portion of a customer's electric service bill creates incentives to conserve energy or choose on-site generation alternatives to avoid the kWh-based charges for not only generation (i.e., energy), but also transmission and distribution.

During the course of retail restructuring, most states have initiated unbundled cost-of-service proceedings to break these traditional two-part tariffs into their respective generation, transmission, and distribution components. One of the more disturbing tendencies in many unbundled cost-of-service proceedings has been the efforts of many UDCs to move much of their distribution-level revenue requirements into fixed, as opposed to variable, charges.

Table 9.3 presents an example of such a proposal. The table outlines the early proposed revenue requirement and rate design for Southern California Edison (SCE) Company in its own unbundled cost-of-service proceeding. What is enlightening is that the table compares the former revenue requirement for distribution under the regime of bundled prices with those proposed for a restructured market. Table 9.3 highlights the dramatic shift from the past recovery of relatively variable-charge-based revenue recovery mechanism to one that is almost exclusively fixed.

Two rate schedules are presented in the table: the General Service 2 (GS-2) and Time-of-Use 8 (TOU-8) rate classes. These two classes typically serve moderately sized and large commercial customers—many of whom have considerable economic opportunities for DER applications. Prior to the revised rate design, SCE recovered 89 percent of its distribution related revenue requirement in the GS-2 class from variable charges, while it recovered 98 percent of that revenue requirement from variable charges in the TOU-8 class.

The shifts to retail competition, however, resulted in a considerable shift in rate-design philosophy by the utility. Under the new rate-design proposals, SCE requested to recover only 31 percent and 44 percent of its distribution revenue requirement from variable charges for the GS-2 and TOU-8 classes, respectively. From a business perspective for UDCs, the change in philosophy makes sense. On a forward-going basis, revenue streams will be considerably more stable since shifts in per-unit consumption associated with

Table 9.3. Proposed Distribution Revenue Requirement Recovery Shift by SCE

	Revenue Requirement			
	Current		Proposed	
GS-2				
Customer Charge	$84,644	11.5%	$248,537	42.9%
Grid Charge			$151,348	26.1%
Total Fixed Charges		11.5%		69.0%
Demand Charge	$221,295	30.0%	$179,631	31.0%
Facilities	$431,469	58.5%		
Total Variable Charges		88.5%		31.0%
T&D Revenue Requirement	$737,408		$579,516	
TOU-8				
Customer Charge	$12,148	2.1%	$8,213	2.7%
Grid Charge			$159,827	53.1%
Total Fixed Charges		2.1%		55.8%
Demand Charge	$309,195	54.7%	$132,846	44.2%
Facilities	$244,378	43.2%		
Total Variable Charges		97.9%		44.2%
T&D Revenue Requirement	$565,721		$300,886	

weather, income, and other factors will have less of a bearing on income than they would under a more variable-based tariff structure.

The implications of this shift in rate design for DER applications is dramatic. Savings under such a distribution-level rate-design formula will be limited to only those savings created through generation. This completely changes the relative economics of a DER project. For example, using a standard sized microturbine, the simple payback under the GS-2 tariff proposed by SCE shifts from 4.8 years to 10.2 years. Under this distribution pricing proposal, a significant capital investment in a microturbine would now take more than twice as long to be paid off.

The above example is just for one utility at a given point in time during the restructuring process, but in many states that have moved forward with restructuring, the theme of fixed-charge revenue recovery is becoming more pervasive. The justification for recovering an increasing proportion of its revenues into fixed charges is based upon a short-run approach to pricing and costs. UDCs argue that their distribution system is relatively static and based for peak-day capacity. As a result, the variable costs of running the system (in the short run) are relatively small; hence, the justification for recovery of the revenues from customer charges and fixed tariff rates.

What this philosophy overlooks, however, is the long-run nature of distribution costs, the incentives for future infrastructure investment, and the benefits that DER could provide in deferring or changing a number of these investments. As noted earlier, optimal rates for distribution service should reflect the long-run incremental costs of providing these services. These long-run costs should include the capacity additions and capital upgrades required to provide distribution service. These additions should figure into distribution rates and send signals to the market about when and where DER is the most economical.

STANDBY AND BACKUP POWER

Standby and backup power is a critically related rate issue that impacts the future development and opportunities for DER. Most recent proposals for DER standby service that are being offered by UDCs do not accurately reflect the lower probability of failure for on-site generation resources and essentially price DER on comparable terms with full-requirements customers (customers who take all of their generation—or their "full requirements—from the utility). In other words, local distribution companies (LDCs) assume that DER customers have a 100 percent probability of on-peak failure and, therefore, these customers should be charged the full-requirements rate. That is, the DER

customer will be expected to pay the same rate that full-requirements customers pay, yet the DER customer may not take any power from the LDC.
Some electric distribution utilities, for instance, impose a grid charge for transmission and distribution to end users that is equal to either the customer peak
demand or the total capacity of on-site generation. Therefore, the LDC prices
the risk of equipment failure at the system peak cost, which should be substantially higher than the true probability of failure. Such a pricing approach
eliminates using DER as a peak-shaving application and essentially limits on-
site generation applications to base-load or nothing, since distribution-level
fees are unavoidable. Unfortunately, these types of rate-design proposals are
being considered and implemented during the course of the overall restructuring process. Such approaches substantially limit the possibilities for DER.

If UDCs had distribution delivery rates that were finely differentiated by time
(system loading), geography, reliability, and volume, then standby charges, as a
separate tariff, would be inconsequential. However, such finely differentiated
rates do not exist, and probably will not exist in the near future. Thus, standby
rates that reflect the different usage characteristics of an end user become one of
the more important rates regulatory commissions can establish for DER applications. Setting these rates for all DER customers at, or closely comparable to,
full-requirements customers is inappropriate and essentially reduces the incentives that end users have for installing DER. Unnecessarily high rates reduce
DER opportunities and customer choice. In many cases, interested customers
could be forced to either walk away from a DER opportunity or develop additional on-site reliability.

As a general proposition, standby charges should reflect the DER customer's choices regarding the level of firm versus nonfirm standby service,
the quantity of standby capacity desired, the location of the customer, and the
probability of using standby capacity during peak constrained times versus
using standby capacity at other times. Regarding the last element, if the
standby rate does not change as a function of time (system loading), then an
estimate of the likely pattern of loading will need to be developed in order to
appropriately structure the standby rate. For instance, one would expect to see
higher standby rates during peak periods in relatively congested dense urban
areas than during off-peak periods in relatively less congested areas.

In terms of firmness, standby service could be differentiated through various degrees of reserved distribution capacity. For instance, customers should
be allowed to enter into agreements with UDCs for distribution service based
on different levels of firm versus nonfirm capacity. Firm distribution-capac-
ity reservations should command a higher price (or premium) than nonfirm
reservations. This type of service differentiation facilitates customer choice
because it allows the customer, and not the UDC, to select the appropriate degree of distribution reliability.

In terms of volume, standby rates should not be set at levels that reflect the maximum peak demand at a particular load location nor the maximum level of installed DER capacity at a particular location as though its use of the distribution system during peak contract time was 100 percent. Standby service needs to reflect the probability of providing emergency power during constrained times versus unconstrained times.

However, standby rates that are set at total installed DER capacity, and at prices similar to full-requirements customers, assume that standby service will be required for a simultaneous failure of all on-site generators. This is a low probability event that would rarely, if ever, happen. Setting rates based on this contingency forces DER customers to pay for capacity that they do not want and would not use. Clearly, this is not a rate-design mechanism that helps facilitate customer choice or economic efficiency.

POSITIVE EXTERNALITIES FROM DER

One of the great advantages of DER is its opportunity to provide a host of additional benefits that go beyond the provision of low-cost, reliable, on-site electric generation. DER can reduce distribution and transmission system upgrades, alleviate generation capacity shortfalls in localized areas, and meet reactive power needs in a particular locale. Thus, DER can contribute in a positive fashion to the electrical grid, and DER owners should be compensated for providing this positive externality. In order to compensate DER providers for their contributions to the system, one needs to be able to determine the value of these contributions. Herein lies the issue: How do you determine a market value for such contributions when no market for these contributions exists? In many cases, these valuation issues, if addressed at all, will be left to regulators, much as they were during the implementation of PURPA-related cogeneration projects. Care should be taken, however, to ensure that these administratively determined solutions do not create serious economic distortions.

From a policy perspective, valuation issues can be broken into two questions. First, what methods should be used to determine the short-run value of DER-provided energy? Second, what methods should be used to determine the longer-run capacity-oriented value of DER? The correct solution to each of these challenges will result in efficient levels of both DER production and capacity. Inappropriately determined methodologies, however, will lead to inefficient levels of DER production, capacity, or both.

The important goal in establishing the short-term value of DER is to ensure that only economic or efficient production occurs. Ensuring that existing rate-design methods do not send inappropriate signals to DER implementation may be the first problem to address. However, of equal importance is the goal of ensuring that the result of too little cost-effective DER does not occur. In order to secure this elusive equilibrium, regulators will have to establish methods for determining short-term value of DER power injected into the grid. Short-term rates, determined through net-metering tariffs or contracts, may send signals to DER providers about the opportunity cost of their power.

A fundamental premise for determining DER buyback or net-metering rates lies in either using existing regional power markets or formulating market prox- ies to send signals about the appropriate levels of efficient small-scale power production. Prices offered for DER-produced power should, at a minimum, re- flect the daily and hourly opportunity cost of dispatching this power into elec- tric power grids. If net-metering rates are settled in this manner, UDCs will ef- fectively serve as aggregators and power marketers for DER power produced within their own grids. This would be a considerable benefit to small-scale DER because most power trades on competitive wholesale markets occur in large (megawatt) blocks. If constructed correctly, the benefits of this aggrega- tion could be extended to UDCs by allowing them a rate of return on their ef- forts to aggregate and market DER power. These themes could even be ex- tended into an incentive regulation application whereby UDCs are allowed to earn higher returns for exceeding certain DER marketing performance targets.

In addition to the relatively straightforward commodity value of DER en- ergy, regulators will also have to consider whether any additional premiums, reflecting the increased value beyond the pure commodity level, will need to be provided. For instance, supplemental premiums for benefits associated with reactive power relief could be included. However, markets for reactive power are exceptionally limited. Additionally, in many instances these mar- kets can be very localized, which leads to the possibility of some market- power problems even if these markets were to arise.

More than likely, regulators will have to establish a proxy for the value of such reactive power, taking care that the valuations are done on a localized and periodic basis. Equally important will be the determination of longer-run capacity-oriented value of DER. Valuations can range widely depending on the locational value of the distributed resource. For instance, the addition of a distributed resource in a high-growth area should be welcome to a UDC. As a result, some value should be assigned to these resources that contribute to alleviating locational problems brought about by high growth. Value deter- mination should include deferred line extensions, deferred substation addi- tions, increased long-run volt-ampere-reactive control, enhanced area load regulation, and reduced system losses.

One method Hoff and Herig (2002) suggest to encourage DER and take ad- vantage of DER positive externalities is to use a tiered pricing structure, at least for distribution costs. Under such a plan, consumers would pay low marginal costs for their initial electricity consumption, and higher marginal costs for greater consumption. In this way, DER could serve as a peak-shaving device.

Stranded Costs

One potential barrier to DER is what many UDCs are describing as distribution-level stranded costs. The argument for the existence and recovery of these stranded costs is similar in nature to those presented in discussions on retail choice, when many high-cost generating assets and long-term contracts were perceived to be in jeopardy. The recovery of stranded distribution costs would prove to be a no-win policy initiative for both DER and competitive retail power markets. UDCs have claimed that DER "strands" distribution assets because investments installed to serve these former full-requirements customers will be no longer be needed. An example of this argument could be portrayed as follows.

Consider a commercial customer that is served by a feeder line that has the capacity to move 10 MW of power. This customer decides to install its own on-site power application of 5 MW. If one assumes that this feeder was fully used, then the installation of the DER application has now resulted in a feeder line that has considerable excess capacity. Now, extrapolate this example to several hundred DER applications of considerable size, as the UDC argument goes, and you have the making of an investment recovery problem. Regulatory commissions, however, should consider a number of countervailing arguments.

First, DER is an incremental technology that is not going to have a large impact like the replacement of utility-owned generation with merchant plants. Massive amounts of localized investments in DER that rival a central-station generation plant, while appealing to many DER providers, are not likely. Therefore, the probability that DER is going to create the same type and magnitude of financial liability that many associate with the past experiences of introducing retail choice is not significant. In other words, this purported stranded distribution cost problem is not likely to result in a multibillion dollar issue such as generation assets did.

Second, most stranded (distribution) cost arguments assume that DER applications will make no contributions to the costs of distribution plant and equipment. This is simply not the case. Appropriate rate-design policies can be developed that send strong signals to DER about their costs on the distribution network, while at the same time sending signals that promote economically efficient applications. Setting stand-by, interruptible, and interconnection rates for DER applications that cover marginal costs and make some contribution to fixed distribution costs actually helps to mitigate the stranded cost problem claimed by UDCs. The goals of rate design for DER should be to encourage, not discourage, grid-connected applications. Yet current UDC pricing practices have just the opposite effect.

Third, it is a questionable proposition as to whether recovery of these distribution assets to UDCs is deprived or simply delayed. Stranded costs have traditionally been defined as the difference between market and book value. These types of costs have arisen in the past because of the transformation of a regulated market to a competitive one. It is important to keep in mind that, given the lumpiness of distribution-level investments, there may be some mismatch between investments and customer growth.

Finally, allowing utilities to recover stranded distribution costs will serve as a perpetual barrier to entry for DER applications and will limit full-scale retail customer choice. As noted above, distribution-level investments are inherently lumpy. Rarely does growth and the discrete sizes for these investments ever match up. Even if they matched, economies of scale in distribution construction should give utilities the incentive to slightly oversize these projects to minimize the number of future upgrades. The inherent lumpiness of investment and the economies of scale in construction imply that there will always be some excess capacity. Thus, UDCs will always have the opportunity to claim stranded costs, unlike past experiences with generation, where these stranded costs, for better or worse, would eventually go away.

Recent Challenges to DER: Reliability, Fuel Availability, and Competition Reliability

One of the leading benefits associated with DER has been reliability. Some of the earliest applications and originally motivating factors for using distributed power applications has been to avoid potential service interruptions that can occur and were growing more frequent in the late 1990s. Two significant events have occurred over the past several years, both of which were catastrophic in nature, challenged DER, and put its potential benefits on the line. In both instances, DER performed in a generally lackluster fashion that exposed one of its most underlying weaknesses: its dependence on a readily available source of fuel.

In the first event, a significant number of installed standby generators in Manhattan failed to start or maintain generation for more than a few hours after the attack on New York City on September 11, 2001. According to a study by Cratty and Fellhoelter (2004), most businesses did not fully comprehend the intricacies of maintenance and operation, and in some places the concentration of dust and smoke quickly clogged air intakes and forced shutdowns.

Perhaps more problematic was the failure to get fuel to these systems. During the 9/11 attack, local distribution gas systems were equally vulnerable and out of service. Diesel-based systems were unable to run, since fuel was at such a high premium during emergency operations, and further, the ability for fuel oil trucks to enter Manhattan was limited. According to reports, fuel oil delivery trucks were allowed into Manhattan only with special escort by the National Guard, prearranged with special authorization (Cratty and Fellhoelter 2004).

The second event, the blackout of August 2003, was on a scale similar to the great 1965 Northeast blackout, which incidentally resulted in the ultimate creation of the North American Electric Reliability Council (NERC). This event did not result in widespread DER usage or subsequent widespread DER implementation. Despite these national catastrophes, DER is not any more pervasive today than it was during the rolling blackouts that originally brought attention to it during 1998 and 1999.

Fuel Availability

The early development of competitive generation markets, both at the wholesale and retail level (i.e., DER), contemplated the availability of relatively cheap and abundant natural gas priced roughly in a range of $2.25 to $2.50 per thousand cubic feet (mcf). No sooner was this assumption made than it began to unwind, as gas prices during the 2000–2001 heating season reached levels that were astronomical relative to historic norms.

As seen in Figure 9.2, gas prices during the 2000–2001 heating season reflected, at least in the near term, a dramatic structural shift. Both the overall

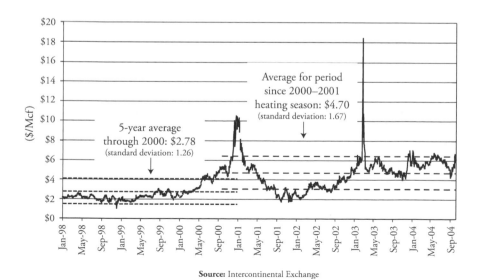

Source: Intercontinental Exchange

Figure 9.2.   Daily Henry Hub Prices 1998–2002

average and volatility of prices have increased since that period and have had important implications for DER, since most are based upon natural gas–fired technologies. If affordable and available fuel supplies cannot be found to run these DER technologies, then all of the issues, and potentially even the solutions, associated with interconnection, rate design, and backup power are esoteric, because the basic economics of most DER applications will be challenged.

The increase in natural gas prices is especially problematic for reciprocating engines and microturbines, both of which run almost exclusively on natural gas. These technologies have been driving the market, and more importantly, the need for new regulations and thinking on DER resources. The pressure that these applications (and their respective developers and vendors) have had on the regulatory process has been a welcome positive externality for other types of DER technologies, like renewables, that have been around for a considerably longer period but failed to get the regulatory traction needed to make any meaningful changes in the way regulators think about DER.

High natural gas prices also impact the more recent darling of DER technologies: hydrogen-powered fuel cells. In his 2003 State of the Union Address, President Bush announced a major initiative in the development of a hydrogen economy. The initiative was precipitated, in part, to move the U.S. economy toward a more green future without adopting the stringent international standards outlined in the Kyoto Protocol. The president noted in his address that the movement toward a hydrogen economy faces a number of considerable obstacles and challenges, not the least of which is the production of hydrogen.

Figures 9.3a and 9.3b present pie charts breaking out the concentration of the various different types of hydrogen production technologies. More than 80 percent of all current hydrogen production comes from steam methane reformation (SMR). This is a well-developed and fully commercialized process for producing hydrogen.

The most common feedstock for SMR is natural gas (i.e., methane). The less commonly used feedstocks in an SMR process are heavier natural gas liquids (ethane, propane, butane) or heavy naphtha. SMR produces a mixture of hydrogen and carbon monoxide that is commonly known as synthesis gas (or "syn gas"). This synthesis gas can be used as a direct fuel (for power generation) or can be separated into its component parts through a shift reactor to make hydrogen. The underlying importance of this process is that the most common method of producing hydrogen (i.e., reformation) is based upon natural gas, and as long as natural gas (and other gas liquids) are priced relatively high, the hydrogen-based economy (along with its commiserate plethora of fuel cells) will be some time in the making.

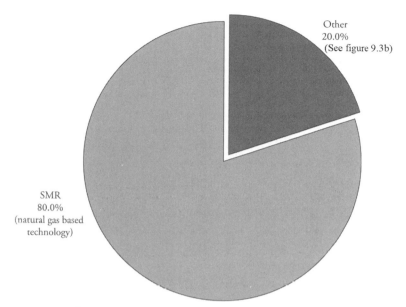

Figure 9.3a.   Hydrogen Production Methods

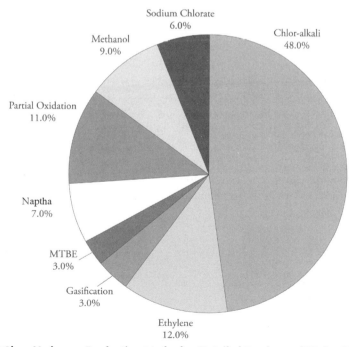

Figure 9.3b.   Hydrogen Production Methods—Detailed Breakout of "Other"

Competition

Another considerable factor dampening the growth opportunities for DER has been the significant growth in highly efficient independent power generation over the past several years. The power outages of the late 1990s put considerable pressure on the industry to develop new physical sources of generation. The industry responded to the generation shortfalls with the announcement of more than 500,000 MW of new capacity—most of which is gas fired, a large portion of which was based on highly efficient advanced combustion turbines in various combined cycle configurations.

Many events in 2001 contributed to the demise of the wholesale power industry including the California market manipulation claims, Enron, a deterioration of the economy (particularly the manufacturing and industrial sectors), the failure of regulation (both state and federal) to complete the process of market reform, and just simple overdevelopment and bad business decisions of several Independent Power Producers (IPPs). The result has been that more than half of the announced generation developments (roughly 225,000 MW) have materialized at the date of this writing.

While wholesale generation development to date has been less than anticipated, it has had an overwhelmingly important impact on wholesale markets, and ultimately retail markets, as more incumbent utilities have been purchasing low-priced spot wholesale market electricity. Prices at both levels have been moving downward due to this increase in competition (and increased overall market efficiency). The increased availability of power, and the decrease in overall market heat rates for power generated in wholesale markets, have driven out a number of opportunities for DER, at least in the short run.

Despite its numerous benefits, DER cannot compete with the laws of thermal dynamics. It is difficult for DER applications, which are paid market prices, to compete with wholesale power that is being generated at heat rates that are considerably lower than their own—especially when other factors discussed earlier are not factored into the rates paid to DER generators. While this problem is transitional in nature and should dissipate as excess generation is whittled away, it does show that DER is not independent of the market. Efficient applications should take their cue from the market, and if more efficient, lower-priced power is available in the market, then it makes no economic sense (from an efficiency perspective) to develop localized generation resources unless there are other mitigating factors that have value and are compared to the strictly lower-cost commodity power.

CONCLUSIONS

DER technology and economics are improving rapidly and being promoted by an increasing number of credible and committed manufacturers. While competition has created a number of opportunities in the energy services business, regulatory policy will still be a pervasive theme conditioning this marketplace. Since DER applications will be tied to the distribution end of the electric power business, regulatory policy will be as important as the economic and technological aspects of these technologies.

DER creates several positive externalities that can aid in the operation of the electricity grid. In addition, one of the attractive features associated with DER is its ability to further the customer choice and retail competition process in the power industry. DER gives customers another service option in meeting their electric power needs. One of the promising applications for the future may very well be with residential customers. Fuel cells, while still in their early stages of development, are a great example of small-scale DER usage.

However, the opportunities for DER will not be created by themselves. There are powerful disincentives for UDCs relative to DER. Leaving the regulatory process to itself, and its traditional stakeholders, will almost guarantee that DER opportunities will be diminished. The UDCs, energy service providers, natural gas industry, DER manufacturers and vendors, and end users all have considerable interests in being engaged in this process with both traditional utility regulators and environmental regulatory interest. The number of DER interests that become engaged in this process during this crucial transition period will determine whether DER will be a common market application or unique niche energy service opportunity.

REFERENCES

Borenstein, Severin (2002). "The Trouble with Electricity Markets: Understanding California's Restructuring Disaster," *Journal of Economic Perspectives* 16(1): 191–211.

Center for the Advancement of Energy Markets, The Role of the Federal Government in Distributed Energy (January, 2002), www.clean-power.com/research/customer PV/Tiered_Rate_Structures.pdf.

Congressional Budget Office, Prospects for Distributed Electricity Generation, September 2003.

Cratty, William, and Fellhoelter, Kevin (2004). "One Year Later: Lessons from the August 14th Blackout Electric Power Grid Independence," *Energy User News* 29 (8): 10–11.

Dismukes, David E., and Kleit, Andrew N. (1999)."Cogeneration and Electric Power Industry Restructuring," *Resource and Energy Economics* 21: 411–29.

Hoff, Thomas E. (1997). "Using Distributed Resources to Manage Risks Caused by Demand Uncertainty," *Energy Journal*, DR Special Issue: 63–84.

Hoff, Thomas E., and Cheney, Matthew (2000). "The Potential Market for Photovoltaics and Other Distributed Resources in Rural Electric Cooperatives," *Energy Journal* 21(3): 113–27.

Hoff, Thomas, and Herig, Christy (2002). "Electricity Rate Structures Can Be Used to Promote Customer-Sited PV: A Lesson From California," www.clean-power.com/research/customerPV/Tiered_Rate_Structures.pdf.

Hunt, Sally (2002). *Making Competition Work in Electricity*. New York: Wiley.

HydroGen, LLC. (2004). The United States Industry Hydrogen Base and Stationary Fuel Cell Power: A Market Analysis. Report Prepared on Behalf of the U.S. Department of Defense Fuel Cell Test and Evaluation Center. (Jefferson Hills, PA: HydroGen, LLC).

Interstate Renewable Energy Council (2004). Database of State Incentives for Renewable Energy, www.dsireusa.org/index.cfm, [October 21].

Lesser, Jonathan A., and Feinstein, Charles D. (2002) "Distributed Generation: Hype vs. Hope." *Public Utility Fortnightly*, June 1.

Woo, Chi-Keung, Horowitz, Ira, and Martin, Jennifer (1998). "Reliability Differentiation of Electricity Transmission," *Journal of Regulatory Economics* 13(3): 277–92.

NOTES

1. Traditional service is defined as a nondifferentiated rate that has a customer charge and constant kWh charge for service. This could be provided by a UDC under its provider of last resort requirements or through a competitive ESP that is challenging utility rates on a per kWh basis and not some differentiated standard like service quality or time-of-use pricing.

2. For a discussion of the cogeneration interconnection requirements associated with PURPA, see Dismukes and Kleit (1999).

3. To work well, interconnection rules must either be so transparent that they can easily be enforced by an outside party or they must be bolstered by incentives that encourage the connectee to cooperate with the connector. It is not an easy task to fulfill either of these conditions. The 1996 Telecommunications Act, for example, required incumbent local exchange carriers (ILECs) to open up their central offices so that competitors could offer services such as DSL (digital subscriber line offering broadband internet services). The ILECs had few incentives to comply, however, and threw up a host of rules and regulations that made connection tedious and difficult.

4. Interstate Renewable Energy Council 2004.

10

Post-Blackout Reliability Policy: A Robust, Flexible Twenty-first-century Grid[1]

L. Lynne Kiesling and Michael Giberson

INTRODUCTION

The blackout of August 14, 2003, affected more than fifty million customers in the U.S. Northeast. The blackout's size and apparently small origin has prompted a lengthy and ongoing review of transmission and reliability policy. Although dramatic, the blackout is just one of many developments that should prompt revision of public policies toward electricity transmission — rapid technological change is bringing new opportunities to improve transmission, and regulatory change to exploit those opportunities should also be considered in any reevaluation of policy.

Technological changes in generation, transmission, distribution, and metering over the past two decades are accelerating the obsolescence of the traditional regulatory model. Changes in other public policies, such as the increasing political support for wind power and other energy sources seen as environmentally friendly, are spurring policy adjustments.[2] Much of the vision guiding recent electric industry restructuring efforts emerged from the deregulatory impulse of the late 1970s, was refined in the 1980s, and became implemented in bits and pieces up through 2000.[3] In the wake of the California restructuring failures and the blackout, opponents of restructuring have been able to stall much of the progress. Proponents of restructuring should take advantage of this pause and period of policy reconsideration to take a broader view of present industry circumstances.

The blackout was a significant event—a costly failure, and worthy of substantial efforts to understand what happened and reduce the probability of additional large blackouts. But policy makers, industry analysts, and citizens/consumers should place what has been learned about the blackout in the context of other industry changes. The promise of industry reform was that it was possible to liberalize regulation and the industry to serve the customer better. This goal is still worth pursuing.

POLICY RECOMMENDATIONS IN THE U.S.-CANADA POWER SYSTEM OUTAGE TASK FORCE FINAL REPORT

The government's final report on the blackout provides crucial background on the operational details of U.S. electric transmission, and it provides a thorough examination of the blackout's causes and consequences.[4] The report offers a snapshot of system conditions on August 14 and a sense of the system dynamics leading up to the blackout. What failed on August 14, 2003, were physical components of the system, but the report rightly focuses on the systemic problems that allowed the failure of a few lines in Ohio to turn into an event that knocked out the power system throughout a good chunk of the northeastern United States and Canada.

The blackout report is a fine example of engineering detective work—the report's authors have done an excellent job of getting the details right—but when it comes to the recommendations, the report misses much of the bigger picture.

A transmission grid that better adapts to unknown and changing conditions would be more resistant to blackouts of any size. The recommendations should help us produce such a system. The kind of rules appropriate to such a dynamic and adaptive system are rules that provide incentives to create local knowledge, improve the use of local knowledge, and provide incentives to use that knowledge in ways that promote overall system performance. Instead, the recommendations propose more rigidity in rules: mandatory reliability standards, more government oversight, penalties for noncompliance, regulatory review of a reliability surcharge to fund an electric reliability organization redesigned by government committee, and other initiatives that tend to centralize control over the transmission system. The main impact of the forty-six recommendations would be the expansion of regulatory oversight over supply-side reliability decision making, not the production, development, and use of local knowledge. Such policies would not enhance system resiliency, flexibility, and adaptability. To achieve those objectives requires a more distributed and decentralized approach.

One aspect of distributed knowledge and control that would increase resiliency is active demand participation in electric power markets. The demand side of the market gets two mentions in the blackout report, in a recommendation calling for additional research funds. Clearly additional research is crucial, since there is still much to be learned about demand-side participation in power markets. However, existing research and programs in place have already demonstrated that consumers as well as producers can contribute to system reliability.

On the supply side, power grid operation and market procedures need to be reexamined for rules that unintentionally generate conflict between private incentives and system reliability. One such conflict was created by the cost-based rules used to pay for reactive power. The blackout report identifies the problem and proposes a change that would eliminate the conflict. Reactive power provides voltage support that enables current to flow on the network, and under existing rules generators owned by utilities or transmission owners receive cost-based payments when they provide reactive power. Independent generators operate under different rules, and may not expect any form of payment for providing reactive power. Current North American Electric Reliability Council (NERC) procedures for Transmission Loading Relief (TLR) are another candidate for this kind of review. TLR procedures are the administrative rules used by transmission operators to control access to transmission lines when the lines become overloaded. The broader organizational framework of transmission also needs reexamination if policy makers hope to resolve the current stalemate in transmission investment.

Our analysis of the recommendations in the blackout report leads to these claims, which form the foundation of our argument:

- Consumer decisions should be allowed to reveal where and when investments in reliability should be made, instead of imposing a potentially inefficient and expensive level of investment.
- Technology and markets allow for the treatment of reliability as a differentiated product, not one uniform level, even though the grid is interconnected. These changes favor markets and contracting institutions over regulatory one-size-fits-all institutions like the ones proposed in the blackout report.
- Although tradition and the blackout report treat reliability as a supply issue, it is also a demand issue. Active demand-side participation in markets inherently acts to moderate strains on the system when system use is properly priced.
- The increasing complexity of the transmission grid increases the need for distributed and decentralized reliability policies adapted to that complexity, and more critically, for policy-making institutions that can adapt to the unknown and changing circumstances in the grid environment.
- Treating grid ownership and management commercially, as a for-profit business, would contribute to obtaining a more flexible and dynamic transmission grid.

The heading in chapter 2 of the task force report reveals the perspective underlying the policy prescriptions: "The North American Power Grid Is One Large, Interconnected Machine."[5] But this "one big machine" picture produces a misconception, or at least an oversimplification, of reality on the grid. The misconception is that reliability on the bulk power grid is an either/or proposition: either it is working, or it is not. The implication is usually that consumers are all bound together by one gigantic externality problem, a problem obviously much too big for markets to resolve. If the one big machine image becomes the framework for public policy, then the answer will be that consumers pay more and get more stringent government oversight of supply-side system reliability.

The transmission grid is not as monolithic as this mechanistic vision suggests. The consumer values created through the transmission grid are the product of thousands, perhaps even millions, of individual decisions: some are large, long-term system-investment decisions and others are small operational and maintenance decisions. These decisions combine to produce a grid that is more reliable in some parts and less reliable in others. Only about 10 percent of the load in the Eastern Interconnection was directly affected by the August 2003 failure, despite the implications of the phrase "One Large, Interconnected Machine." Learning what failed on August 14, 2003, is important: which wires, which policies, and which actions proved to be insufficient. But it is also important to learn what succeeded—where was the grid flexible enough, where were sufficient policies in place, where did system operators take the right actions.

DISCOVERING THE RIGHT AMOUNT OF RELIABILITY

The knee-jerk policy reaction, and indeed the presumption in the blackout report, is that electric power consumers need a more reliable transmission grid. The truth may be, on the other hand, that consumers may be better off with less reliability. As the blackout report points out, large blackouts are extremely uncommon. The electric power system is already highly reliable, but every ten years or so a significant blackout occurs.

Large blackouts are costly, and "never again" is fine rhetoric, but do consumers really want more reliability? Consumers already pay a great deal for reliability, and whether they want to buy more depends in large part upon how much will it cost. Reliability has been treated as an engineering matter, but it needs to be treated as a customer service and as part of a commercial product. The task force report has offered a long shopping list of top-down recommendations, but it did not include a price tag.

Reliability Is a Supply *and Demand* Issue

A blackout appears to be a supply failure, so the natural impulse is to look for supply-side solutions: more transmission lines, high-tech system monitoring, building power plants closer to population centers, better grid planning and testing procedures. But on the demand side of the market, consumers can help make the grid more reliable by becoming more engaged in the market. Consumer participation in markets enhances reliability because it harnesses downward-sloping demand and results in customers decreasing quantity demanded precisely in hours and at places in which the system is facing the highest strain. New technologies for metering and end-user voltage management are allowing the demand side to make more active contributions to managing reliability as well.[6]

The decentralized and distributed network of customers can contribute to grid resiliency and flexibility through more active participation in the market. Active, engaged customers could choose anything from a fixed price that incorporates an insurance premium to full real-time pricing, in which the customers bear the financial risk of price volatility, and they could see electricity bills fall by shifting or reducing use.[7] Several ongoing demand-response and retail-pricing pilots have demonstrated that even residential customers do shift their demand away from peak in response to prices. Even if the magnitude of the shift is small, the effect may be large because of the nonlinear relationship between peak load reductions and network reliability. A small load reduction at just the right time can keep the system from hitting capacity.

Technological developments give consumers another tool for managing their energy use. Consumers can set electricity monitors to increase air conditioning temperatures if prices go above a certain amount, for example, or can shift manufacturing schedules to minimize electricity use during peak hours. Right now, with almost all (even many industrial and commercial) U.S. consumers paying average prices, consumers have little incentive to manage their consumption and shift it away from peak hours during the day.

Reducing peak use also contributes to greater operational security, as fewer reserves are necessary to maintain reliability, and eases stress on adequacy planning, as the need for system expansion to support ever-greater system peak loads is diminished.[8] Both historical experience and laboratory experiments show that electricity customers do respond to price changes, and that both suppliers and customers can be better off from doing so.[9] This option does not currently exist for most customers in most places.

Another approach to enabling consumers to contribute directly to reliability comes from efforts to turn demand response into a tool that transmission system operators can call on in their efforts to keep supply and demand constantly in balance. In this vein, responsive customer demand can be used as spinning reserves dispatched by the system operator to meet system reliability requirements. For example, a significant portion of the California Independent System Operator's spinning reserve requirement could be supplied from the California Department of Water Resources (CDWR) pumping load. Given the appropriate economic incentives, the CDWR could stop pumps for brief intervals to make small adjustments in response to specific short-term transmission system needs. Another approach would enable controllable air conditioning units to be cycled off for brief periods when the system is stressed.[10]

Reliability as a Differentiated Product

There is another barrier to efforts to give the public the level of reliability that they want. It is not clear that anyone knows whether a typical electricity consumer in the United States or Canada would rather have more reliability at higher power prices or less reliability and lower power prices. More to the point, some consumers would probably be willing to pay more to have more reliable service, and others would choose lower prices even if service quality went down a little. The existing top-down system offers few avenues for suppliers, operators, or policy makers to gather information on how different customers value reliability. Nor does the regulatory system offer many opportunities to energy consumers wishing to buy different levels of reliability; instead industry and its regulators tend to see reliability as a one-size-fits-all characteristic.[11]

If reliability is defined as the local electric company being willing and able to supply uninterrupted electric power at a price and on demand, it is possible for neighboring customers to have differing reliability. An industrial customer may be willing to be cut back on power consumption when the system is stressed, contractually agreeing that the electricity provider can reduce service under certain circumstances. The customer is in effect accepting lower local reliability to allow the electric company to provide added reliability to other customers.

But this interruptibility option still embodies a supply-side view of reliability. If reliability is defined as the uninterrupted provision of services to the consumer from electric appliances, then we can see even more clearly the limitations of the one-size-fits-all view of reliability. Consumers who have specific needs for reliable power can, for example, purchase battery-backup power supplies to help keep computers up and running even if the local power company is having problems with delivery: these consumers are paying a little more to have more reliable service from a select appliance or two, and as a result would feel less of a need for uninterrupted electric power services from the local power company. Consumers also may have similar tolerances for service interruptions, for instance, being able to tolerate periods of having their air conditioners turned off during the summer. Devices are available that would enable a consumer or third-party energy management company to control air conditioner loads remotely. Businesses do the same thing on a larger scale, with companies that have special needs for highly reliable electric power spending millions of dollars to secure their supplies, and similarly can install complex energy-management systems to control power consumption. This is a focused demand-side approach that provides very targeted power reliability.

These consumer-side choices have the potential to reveal more information about consumer values for reliable electric power system service, but current approaches to providing targeted reliability services usually prevent the local utility from being directly involved. As a result, few avenues exist for consumer reliability choices to percolate up through the market informing the systemwide choices about reliability that distribution and transmission system operators need to make appropriate maintenance and investment decisions. Chao and Wilson (1987) have suggested a "priority insurance" system that would produce this kind of information.[12]

The essence of priority insurance is to have the distribution company pay consumers when the lights go out. A simple idea, but Chao and Wilson add a twist: the electric company offers different qualities of service. For a higher price, one can obtain a lower probability of being cut off when the system is short of power (and a higher payment from the electric company when the lights go out); pay a lower price, get a higher probability of being cut off (and a lower payment). Customers would be able to choose between price and reliability. When the embedded premiums are set appropriately, each customer class can be assured of being no worse off than before, and they may be made better off.[13] Noussair and Porter (1992) experimentally tested a version of this idea against a simple system of proportional rationing of shortages, and found that their version of priority insurance was more efficient.[14]

While one benefit of the priority insurance approach is that it allows the electric company to allocate a shortage efficiently by having customers prioritize their own use, a larger payoff comes from the information created by the consumer actions. Priority insurance allows consumers to evaluate energy and service reliability separately, allowing the company to distinguish what customers are willing to pay for power from what they are willing to pay for reliability. Companies could target investments where they provide the most long-term value to consumers. This information about how customers value reliability is crucial to getting the efficient amount and kind of system infrastructure investment.

Consumers are the sleeping giant of electric reliability. Retail rate regulation has put the demand side to sleep, but it is time for consumers to wake up. Retail pricing, including the separate pricing of energy and reliability, is a crucial component of a healthy, dynamic electricity industry, and a reliable grid. Offering consumers service choices in a range of prices would make diverse consumers better off, and bolster system reliability.

THE ROLE OF TRANSMISSION OWNERSHIP STRUCTURE[15]

Achieving a more robust and reliable grid requires reconsideration of policies toward ownership, management, and operation of transmission. The present organization of transmission operations is complicated and not particularly transparent.[16] Currently most transmission is owned by regulated public utilities, though federal agencies and other entities own a large chunk of transmission, especially in the West. In the northeastern United States, the Midwest, and in California, one of several independent transmission providers (RTOs or ISOs) manages transmission, while in the rest of the United States local monopoly utilities and federal agencies manage most transmission.[17]

The bulk of transmission service is formally regulated by the Federal Energy Regulatory Commission (FERC), a federal government agency. A great deal of transmission was built by local monopoly utility companies to serve ratepayers in their "home" service territory, under terms regulated by state utility commissions. Even while regulated transmission rates are set at the federal level, however, a good deal of the revenue that pays those rates comes from state-regulated retail electric rates. As a result, state policies retain influence over transmission policies.

Most reliability rules, governing a great deal of the terms of transmission operation and the costs involved, are established by the North American Electricity Reliability Council (NERC) and implemented in conjunction with twelve regional reliability councils. In regions with RTOs/ISOs, that organization usually acts as reliability coordinator, overseeing control area operators. The control area operators are the "front line" system operators with the job of keeping the interconnected grid up and running.

The Midwest region, which saw the origin of the blackout, is much more complicated than most of the country. The explanation of system operation and control in the Midwest takes an entire half-page sidebar on page 14 of the blackout report. In the Northeast, the ISOs typically cover one or two control areas; at the time of the blackout the Midwest ISO provided "reliability coordination for 35 control areas in the ECAR, MAIN and MAPP regions and 2 others in the SPP region." PJM Interconnection, the transmission system operator in the Middle Atlantic states, was originally a single control area, but because of expansion into western Pennsylvania and the Midwest now oversees nine control areas.

This complex organizational structure to control reliability arose out of the 1965 blackout, which occurred at a time when wholesale power transactions were few, and little trade crossed control area lines. Now, with power flows crossing borders between reliability coordinators and through multiple control areas, things have changed, but the tools available to manage reliability have not kept up. At the time of the blackout, the Midwest ISO had oversight authority for some, but not all, of the reliability functions in the region. The tools available to the Midwest ISO—for example, ordering transmission loading relief (TLRs), which requires the ISO to contact multiple control areas to coordinate system adjustments—were clumsy at best.

Another issue in transmission policy and reliability has been whether sufficient investment incentives exist. Investment in transmission has been lagging for years, and the regulatory response has been to offer more incentives and more assurances that cost recovery is available.[18] In April 2004, FERC issued a policy statement on reliability that again assured transmission owners that prudent reliability costs could be passed along in regulated transmission rates.[19] Cost recovery is more of the same regulatory approach; maybe this time it will work. Or, maybe, the problem is not mainly one of uncertainty about full recovery of prudent expenditures.

Simplifying the organizational structure surrounding reliability policy, or more accurately, *commercializing* reliability policy, would help enable investment in transmission. Traditional NERC reliability principles have been implemented as voluntary standards. The blackout report has called for reliability standards that are enforceable and backed by regulatory authorities. These approaches will do nothing to reduce the current complexity surrounding the management of reliability. Commercializing reliability—making it a matter of contractual relationships between interconnecting transmission owners and the generators and distribution utilities that connect to the grid—could bring more clarity and certainty to the process of investing in reliability than yet another regulatory policy statement.

A similar perspective is developed by Kleindorfer (2004), who points out that existing rules governing transmission ownership and operation do not provide the transparency and clarity necessary to any commercial ventures. "The urgent matter of providing incentives for coordinated resolution of interdependent reliability and congestion problems," he said, "will remain unresolved until we move from the autarchic perspective of 'every man for himself' to the view that emphasizes the need to see transmission service provision as a business."

Kleindorfer's discussion focuses on four commercial principles that he argues would make transmission a forward-looking venture that would attract investment in, among other things, reliability. First, transmission entities (call them RTOs for brevity) have to face performance standards and to be accountable for their achievements and failures. This accountability is the role that capital markets and shareholders play in for-profit companies. Second, RTOs should focus on customers. Third, operations and planning in RTOs must integrate the engineering of the system with its economics. Finally, the RTO governance structure must be responsive and decisive. Through the transparent property rights structure and incentives that for-profit commercial ventures face, capital markets could see more potential value, and see it more clearly, in transmission investments.

Unfortunately, FERC has chosen a regulatory rather than a market approach. In the 1990s FERC began considering policy changes that would enable transmission to evolve to keep up with the changes in the generation sector. FERC was concerned that a lack of transmission infrastructure investment would stifle generation markets. To that end FERC passed a series of regulations to promote open access to the transmission grid on fair and reasonable terms. Throughout these regulatory changes, investment in transmission has been backed up by regulatory guarantees of cost recovery. Yet transmission ownership structure remains fragmented and investment remains largely stagnant.

Underlying structural ownership issues may not just be vague, but even actively harmful: incumbent transmission owners may face economic incentives contrary to overall system quality and performance.[20] Writing about the economics of networks, Cremer, Rey, and Tirole (2000) make the essential point in a different context: the benefits of network quality improvements may go disproportionately to the creative upstarts in the industry, but the quality of the network is largely determined by the investment decisions of larger, established firms.[21] If you are the established firm, how much do you want to pay in order to throw the door open wide to your new competitors?

The current regulatory/administrative approach to transmission planning and operations has, along with a substantial dose of regulatory uncertainty, given us the current mess in the transmission business. The solution may be to treat the transmission business as more of a business.

A TRIP TO THE ISLANDS

The discussion so far has focused on regulatory, market-based, and organizational changes that could help consumers get more value from the transmission grid. Additional research in good old-fashioned power systems engineering is necessary as well. Changes in the patterns of use of the transmission grid require fresh looks at old approaches, and cutting-edge research seeks to contribute new tools for understanding and responding to stress on the transmission grid.

One way to make the grid work better for consumers is to figure out how it can fall apart more gracefully. The ability of a subregion to cut itself off from the surrounding grid—islanding—can help limit the cascading failures that constitute widespread blackouts. But islanding can also cause problems of its own, if not done well. If the subregion that separates is itself not already in approximate balance between generation and load, it may also fail. In addition, the system left behind by the islanders may become even more difficult to manage as subregions drop out.

The details of the cascade in chapter 6 of the Task Force final report indicate that at 4:10:39 P.M., the tripping of protective relays isolated Cleveland and Toledo from the rest of the grid. This island was unstable because there was insufficient generation inside the island. Thus, the island dropped load but could not stabilize and blacked out. At the same time, Detroit suddenly had a great deal of excess power because the Cleveland and Toledo load separated from the system, and this excess power bounced back through the system tripping additional relays and leading to the separation of much of the northeastern United States and Eastern Canada from the rest of the eastern interconnection. At the same time, most of New England and the Maritimes separated from the surrounding grid in the East and kept operating, as did small regions in New York, Ontario, and Quebec.

Research into dynamic islanding has looked for ways for the system to "fall apart more gracefully."[22] The idea behind dynamic islanding is that the system could be intentionally, endogenously separated into self-sustaining parts—based upon current system operating conditions—rather than allowing haphazard islanding to result from individual equipment trips. The islanding on August 14 was the result of individual relay trips, not coordinated responses, and did little to stabilize the system.

The exploration of dynamic islanding is simply part of the process of understanding how the transmission grid is changing. The partial restructuring of the industry that the country has gone through over the past few years has led to significant changes in the use of the grid. In the regions in which wholesale restructuring has most advanced, in RTOs with central coordination of the regional transmission system integrated with power markets, use of generation and transmission resources are carefully coordinated so as to avoid placing the system at risk. In principle, the markets and operating systems that the Midwest ISO is putting into place should give them similar protection once their markets start up.[23] But much of the country operates outside of such regional operations, and patterns of use are changing outside of RTOs as well.

In another approach to understanding the changing dynamics of the transmission grid, research combining network theory with power systems engineering is providing additional insights into system reliability and cascading failures.[24] A statistical examination of NERC blackout data by Carreras and coauthors (2004) showed that larger blackouts were more frequent than expected under standard assumptions about how blackouts spread. After additional research, one of their surprising insights was that local efforts to prevent small blackouts may cause the system to become more susceptible to larger blackouts.[25] Each local effort to boost reliability enabled the local system operators to feel more comfortable operating with a narrower "safety margin," but the combined effect was to reduce the resilience of the overall system. With narrower safety margins in play, there is less slack in the system, and one small failure can more easily propagate throughout the network.

The blackout report did make some recommendations related to complexity, cascading, and islands. It seeks additional support for research, and suggests a number of physical and cyber security changes. Such research can help reduce the costs and risks of blackouts without scaling back on the use of the transmission grid to support competitive wholesale power markets.

THE ROLE OF VOLTAGE MANAGEMENT AND REACTIVE POWER

On the supply side, power grid operation and market procedures need to be reexamined for rules that unintentionally create conflict between private incentives and system reliability. The blackout report urges that "[m]arket mechanisms should be used where possible, but in circumstances in which reliability and commercial objectives cannot be reconciled, they must be resolved in favor of high reliability."[26] Clearly a transmission system operator must at some time invoke emergency rules to maintain the short-term reliability of the grid, and market rules reasonably are suspended during the emergency. But the premature granting of a trump card to high reliability in cases of conflicts overlooks how poorly designed reliability rules can create the conflicts in the first place.

One example of how transmission system rules can create conflict was identified in the blackout report—incentives to provide reactive power. In alternating-current electric networks, reactive power contributes to the maintenance of a magnetic field sufficient to keep network voltage within a particular range. Adequate provision of reactive power services is critical to the reliable operation of the transmission system. Reactive power cannot actually transfer energy or do any work itself, but it enables current to flow and to transfer energy from producer to consumer. Without adequate reactive power, voltages on the transmission system can drop and the system can become unstable. In extreme cases, insufficient reactive power levels result in the system's failure to deliver real power.

The failure to maintain adequate reactive power support was cited by the Task Force *Final Report* as among the direct causes and contributing factors leading to the blackout that originated in the Midwest. Reportedly, system operators asked an independent generator to produce more reactive power, and the generator declined because it would make more profit selling real power.[27] The general problem is that no matter how valuable reactive power becomes to the transmission system, in most of the country the prices that the transmission provider can pay for reactive power remain fixed in cost-based rate tariffs on file at the FERC.[28] When real power prices rise to high levels, the rules governing reactive power purchases create conflict.

The blackout report rightly recommends that when generators are called upon to sacrifice real power sales to provide reactive power for reliability purposes, the generators should be paid for any lost revenues. The blackout report specifically called upon FERC and appropriate Canadian authorities to require all tariffs or contracts for the sale of generation to include provisions specifying that the generators can be called upon to provide or increase reactive power output if needed for reliability purposes, and that the generators be paid for any associated lost revenues. In other words, the blackout report recommended that generators providing reactive power services to support the network be paid for the opportunity cost of their provision of reactive power. That change would remove the conflict between private incentives and system reliability.

Current NERC procedures for Transmission Loading Relief (TLR) are another candidate for this kind of review.[29] TLR procedures provide a set of administrative rules to cut back power transactions when congestion threatens to overload part of the grid. TLR rules are known to be economically inefficient at resolving congestion, and often are even inefficient in a technical engineering sense because the transactions that get cut off sometimes offer only slight relief of the problem.[30] The expectation that a TLR may be called can create perverse incentives for transactions in the neighborhood of the expected TLR. But the economics surrounding TLRs are murky, and a deeper examination is needed.

In some limited circumstances, concern for system reliability should trump market decision making, but poorly designed reliability rules should not be the source of the problem.

CONCLUSION

The most obvious lesson learned from the blackout is that the electric power industry and its regulatory organizations are better at diagnosing system failure *ex post* than at divining ways to foster growth of a self-correcting, self-reinforcing, and dynamically reliable system. The blackout report did an excellent job of diagnosing the recent failure, but the forty-six recommendations it offers will not support the growth of a robust, dynamic, flexible grid.

Many of the relevant factors that add to or subtract from system reliability are well understood. Most of these factors are attributed to, or could be measured and attributed to, the responsible party. The responsible party could then be either charged or paid an appropriate amount. The key is to bring reliability into the commercial realm, where choices can be made in the presence of relevant tradeoffs.

The ultimate objective is healthy, thriving wholesale power markets, and a healthy wholesale market requires a robust transmission network. Reliability is a crucial element in enabling those power markets to continue developing, but that does not mean that reliability is a one-size-fits-all characteristic of the network. Treating reliability as a public good leads to conflicts; treating it as a private good could avoid those conflicts. The metering, monitoring, and switching technology exists to treat reliability as a differentiated, private good. Now the institutional and legal structure must adapt and evolve to take advantage of these opportunities.

Healthy electricity markets are a pipe dream without a demand side. Active retail demand transmits end-customer preferences into the wholesale market, smoothing out peaks and optimizing load factors (and curbing the exercise of supplier market power along the way). Furthermore, allowing demand reduction to bid into capacity markets can reduce the construction of new generation and transmission capacity, and is therefore a good long-run strategy for conservation of resources and for making investment more efficient. Demand response is the Swiss army knife of the electricity policy world—it is one compact tool that does a lot of things in a very parsimonious way.

In the end, the most important changes to make in the industry are really just a continuation of industry restructuring. Let us commercialize reliability—reform the reliability rules to align incentives and information flows. Reliability is valuable to consumers. What has been lacking is a way for consumers to express those values, and for suppliers to be paid appropriately for providing it.

The introduction to the Task Force's chapter of recommendations is telling both in its focus—fixated on the supply side—and its proposed measure for success. The report provides four broad themes to use in thinking about the recommendations. Distilled to their essence the four themes are:

- Regulators and industry must commit themselves to the highest reliability standards in "the planning, design, and operation of North America's vast bulk power system." Market mechanisms "should be used where possible," but not if they would conflict with reliability.
- High reliability is costly, regulated firms must be assured of cost recovery, and unregulated firms must believe that reliability investments will be profitable.
- Recommendations must be implemented to work, and industry and government should commit themselves to the task.
- While the August 14 blackout was not caused by a malicious act, a number of physical and computer security improvements are called for.[31]

The demand side of the equation, consumers, gets explicit mention only in theme number two. The report tells consumers (and regulators) that reliability is not free. Ironically, it is the same message that consumers must tell the Task Force: reliability is not free. Before policy makers commit those they represent to the "highest reliability standards," consumers, ratepayers, and taxpayers should ask how much it would cost. Better still, what is the value on the margin: How much more reliability will we get as spending on reliability increases?

Another important demand-side question is how customers will know if they get the level of reliability that they pay for. The Task Force proposes to measure the success of its program by how many of its proposals get implemented: "The metric for gauging achievement of this goal," says the report, "will be the degree of compliance obtained with the recommendations presented below." It is typical of the supply-side focus of the Task Force, trying to gauge the value of a process by the input into the effort. The consumer, who will be stuck with the bill, cares more about how well the resulting system works.

The regulator's report, and the regulatory recommendations and actions, provided a regulatory solution. That fact in itself is not too surprising. But the regulatory solution will not give consumers what they could really use, which is a more efficient, more resilient, more adaptable, and more dynamic power grid.

REFERENCES

Albert, R., I. Albert, and G. Nakarado. (2004) "Structural Vulnerability of the North American Power Grid," *Physical Review E* 69: 025103(R).

Alvarado, Fernando, and Rajesh Rajamaran. (2000) "On the Inherent Inefficiencies of TLR for Trading Electricity," Presentation, MEET 2000, Stanford, CA.

Aviram, Amitai. (2003) "Regulation by Networks," *BYU Law Journal* 2003: 1179–238.

Black, Jason. (2004) "Demand Response as a Substitute for Electric Power System Infrastructure Investments," Electricity Transmission in Deregulated Markets: Challenges, Opportunities, and Necessary R&D Agenda. Carnegie-Mellon University (Pittsburgh, PA). Working paper available at www.ece.cmu.edu/~tanddconf_2004/BlackJ%20CMU%20Paper.pdf.

Carreras, B., V. Lynch, D. Newman, and I. Dobson. (2003) "Blackout Mitigation Assessment in Power Transmission Systems," Thirty-Sixth Hawaii International Conference on System Sciences. Available at eceserv0.ece.wisc.edu/~dobson/PAPERS/complexsystemsresearch.html.

Carreras, B., D. E. Newman, I. Dobson, and A. B. Poole. (2004) "Evidence for Self-organized Criticality in a Time Series of Electric Power System Blackouts," *IEEE Transactions on Circuits and Systems I*, 51: 1733–40.

Chao, Hung-Po, and Robert Wilson. (1987) "Priority Service: Pricing, Investment, and Market Organization," *American Economic Review* 77: 899–916.

Cremer, Jacques, Patrick Rey, and Jean Tirole. (2000) "Connectivity in the Commercial Internet," *Journal of Industrial Economics* 48: 433–72.

Eto, Joe, et al. (2001) *Scoping Study on Trends in the Economic Value of Electricity Reliability to the U.S. Economy*, LBNL-47911, Lawrence Berkeley National Laboratory (Berkeley, CA).

Federal Energy Regulatory Commission. (2004a) *Policy Statement on Matters Related to Bulk Power System Reliability*, 107 FERC 61,052.

———. (2004b) *Assessing the State of Wind Energy in Wholesale Electricity Markets*, FERC Staff Briefing Paper, Docket No. AD04-13-000.

———. (2005a) *Interconnection for Wind Energy and Other Alternative Technologies*, Notice of Proposed Rulemaking, Docket no. RM05-4-000.

———. (2005b) Principles for Efficient and Reliable Reactive Power Supply and Consumption, FERC Staff Report, Docket No. AD05-1-000.

Fumagalli, E., J. Black, I. Vogelsang, and M. Ilic. (2004) "Quality of Service Provision in Electric Power Distribution Systems through Reliability Insurance," *IEEE Transactions on Power Systems* 19(3): 1286–293.

Giberson, Michael, and Lynne Kiesling. (2004a) "Analyzing the Blackout Report's Recommendations: Alternatives for a Flexible, Dynamic Grid," *Electricity Journal* 16 (July): 51–60.

———. (2004b) *Mercatus Center Public Interest Comment on Midwest ISO Proposal Concerning Reactive Power Procurement*. Available at http://www.mercatus.org/pdf/materials/811.pdf.

Hamachi LaCommare, Kristina, and Joseph H. Eto. (2004) *Understanding the Cost of Power Interruptions to U.S. Electricity Consumers*, LBNL-55718, Lawrence Berkeley National Laboratory (Berkeley, CA).

Hirst, Eric. (2002a) "The Financial and Physical Insurance Benefits of Price-Responsive Demand," *Electricity Journal* 15(4): 66–73.

———. (2002b) "Reliability Benefits of Price-Responsive Demand," *IEEE Power Engineering Review* 22(11): 16–21.

———. (2004) *U.S. Transmission Capacity: Present Status and Future Prospects*, Edison Electric Institute and U.S. Department of Energy, Washington, DC, June. http://www.ehirst.com/PDF/TransmissionCapacityFinal.pdf.

Hirst, Eric, and Brendan Kirby. (2001) *Transmission Planning for a Restructuring U.S. Electricity Industry*, Edison Electric Institute, Washington, DC. www.ehirst .com/PDF/TransPlanning.pdf.

Hunter, Richard, Ronen Melnik, and Leonard Senni. (2003) "What Power Consumers Want," *The McKinsey Quarterly* 2003 Number 3.

Joskow, Paul. (2005) "Transmission Policy in the United States," *Utilities Policy* 13:95–115.

Joskow, Paul, and Richard Schmalensee. (1983) *Markets for Power: An Analysis of Electric Utility Deregulation*, (London, England, and Cambridge, MA: MIT Press).

Kiesling, Lynne, and Adrian Moore. (2003) *Movin' Juice: Making Electricity Transmission More Competitive*. Reason Public Policy Institute Policy Study 314. Available at www.rppi.org/ps314.pdf.

Kirby, Brendan, and John Kueck. (2003) *Spinning Reserve from Pump Load: A Technical Findings Report to the California Department of Water Resources*, ORNL-TM/2003/99, Oak Ridge National Laboratory (Oak Ridge, TN).

Kleindorfer, Paul. (2004) "Economic Regulation Under Distributed Ownership: The Case of Electric Power Transmission." Available at www.charlesriverresearch-corp.com/Kleindorfer.pdf.

Newman, D. E., B. A. Carreras, V. E. Lynch, and I. Dobson. (2004) "The Impact of Various Upgrade Strategies on the Long-term Dynamics and Robustness of the Transmission Grid," Electricity Transmission in Deregulated Markets: Challenges, Opportunities, and Necessary R&D Agenda. Carnegie-Mellon University (Pittsburgh, PA). Working paper available at eceserv0.ece.wisc.edu/%7Edobson/PAPERS/newmanCMU04.pdf.

Noussair, Charles, and David Porter. (1992) "Allocating Priority with Auctions: An Experimental Analysis," *Journal of Economic Behavior and Organization* 19(2): 169–95.

Potomac Economics. (2003) *2002 State of the Markets Report: Midwest ISO*.

Rassenti, Stephen, Vernon Smith, and Bart Wilson. (2002) "Using Experiments to Inform the Privatization/Deregulation Movement in Electricity," *Cato Journal* 21: 515–45.

———. (2003) "Controlling Market Power and Price Spikes in Electricity Markets: Demand-Side Bidding," *Proceedings of the National Academy of Science* 100: 2998–3003.

Reiss, Peter C., and Matthew W. White. (2003) *Demand and Pricing in Electricity Markets: Evidence from San Diego During California's Energy Crisis*, NBER Working Paper No. 9986, Available online at http://www.nber.org/papers/w9986.

Schweppe, Fred C., Michael C. Caramanis, Richard D. Tabors, and Roger E. Bohn, (1988) *Spot Pricing of Electricity* (Boston, MA: Kluwer Academic Publishers).

Smith, Vernon, and Lynne Kiesling. (2003) "Demand, Not Supply," *Wall Street Journal*, August 20, 2003.

U.S.-Canada Power System Outage Task Force. (2004) *Final Report*. Available at https://reports.energy.gov.

You, H., V. Vittal, and Z. Yang. (2003) "Self-Healing in Power Systems: An Approach Using Islanding and Rate of Frequency Decline-Based Load Shedding," *IEEE Transactions on Power Systems* 18: 174–81.

NOTES

1. This chapter is an extension of the arguments presented in Giberson and Kiesling (2004a). We are grateful to Rob Gramlich and Ed Reid for initial comments that clarified our thinking, and to Andrew Kleit and Alex Tabarrok for helpful comments and editing.

2. See FERC Notice of Proposed Rulemaking, Docket No. RM05-4-000, *Interconnection for Wind Energy and Other Alternative Technologies* (January 2005); FERC Staff Briefing Paper, *Assessing the State of Wind Energy in Wholesale Electricity Markets*, Docket No. AD04-13-000, (November 2004).

3. In different ways the policy-oriented work of Joskow and Schmalensee (1983) and the technical work of Schweppe, Caramanis, Tabors, and Bohn (1988) contributed to the vision that has guided restructuring the industry.

4. U.S.-Canada Power System Outage Task Force, *Final Report*, available at https://reports.energy.gov. (Henceforth referred to as the Task Force *Final Report*, or simply as the blackout report.)

5. Task Force *Final Report*, p. 6.

6. The Task Force recommendations do address demand-side issues in a small way. In two short paragraphs in a discussion of research needs, the report: (1) cites "demand response initiatives to slow or halt voltage collapse" as one aspect of research into ways to prevent cascading power outages; and (2) urges the "study of obstacles to the economic deployment of demand response capability and distributed generation" (Recommendation 13: DOE should expand its research programs on reliability-related tools and technologies. Task Force *Final Report*, page 149.) In addition, with a little creativity, a role for demand response might be read into a few other recommendations. For the most part, however, the blackout report is about a supply failure and supply-side proposals. For most of the blackout report, end-use consumers are simply the "load," a passive burden that the supply side must simply bear.

7. Hirst (2002a).

8. Hirst (2002b); Smith and Kiesling (2003); Black (2004).

9. See Reiss and White (2003); Rassenti, Smith, and Wilson (2002, 2003).

10. Kirby and Kueck (2003). The research was supported in part by the Department of Energy's current research program in transmission reliability.

11. Hunter, Melnik, and Senni (2003) suggest that consumers that have relatively reliable power would rather see improvements in call center waiting times, and other service improvements, rather than still more reliable power. Hamachi LaCommare and Eto (2004) and Eto et al. (2001) provide assessments of the value of reliability.

12. See also Fumagalli et al. (2004).

13. Chao and Wilson (1987) describe how to set the rates to ensure this result. The significant question is empirical: How much better off can consumers become, and how certain can we be that consumers will be sufficiently better off?

14. Noussair and Porter (1992) added another twist in that in their version the number of levels of service and price levels were endogenous to the customer evaluations.

15. For additional detail on the development of current transmission policy in the United States, see Joskow (2004).

16. Regional Transmission Organizations (RTOs) and Independent System Operators (ISOs) enjoy slightly different status within FERC policies, but for the most part the distinctions are not relevant to our discussion here, and we will refer to them generically as RTOs. See the discussion of Order No. 2000 below.

17. Hirst (2004), Hirst and Kirby (2001).

18. FERC (2004a).

19. Paul Kleindorfer, "RTO Transco Design." Presentation at *The Future of Electricity Policy*, a Progress & Freedom Foundation Workshop, April 5, 2004. Available at www.pff.org/publications/energy/.

20. See also Aviram (2003).

21. See, for example, You, Vittal, and Zang (2003).

22. The Midwest ISO market start-up was planned for April 1, 2005.

23. Carreras et al. (2004). Several related papers are available online at eceserv0 .ece.wisc.edu/~dobson/PAPERS/complexsystemsresearch.html. Some of this work has been supported by DOE's transmission research effort.

24. See Carreras et al. (2003), Newman et al. (2004).

25. Task Force *Final Report*, p. 139.

26. Federal Energy Regulatory Commission (2005b).

27. Rules in the Midwest ISO at the time of the blackout were even worse than typical. The Midwest ISO tariff did not provide a means to directly pay generators for reactive power, but rather paid control area operators (CAO) for providing services. While CAO tariffs filed with FERC would include a revenue requirement for costs associated with reactive power capability of affiliated generators, most such tariffs did not include any compensation mechanism for generators not affiliated with the vertically integrated utility. The ISO has recently proposed changes to its tariff to accommodate payments to all generators on comparable terms. See filings in FERC docket ER04-961-000, including Giberson and Kiesling (2004b).

28. See *NERC Policy 9 – Reliability Coordinator Procedures*.

29. See Alvarado and Rajaraman (2000) and Potomac Economics (2003).

30. Task Force *Final Report*, p. 139.

Index

access charges, 151–53
Agent. *See* Independent System
Operator (ISO)
Aigner, D., 49
Alberta August 2000 PPA Auction
Results, 103–4, *103*
Alberta Department of Energy, 104
Alberta Interconnected Electricity
System availability capacity, *96*, 97
Alberta post-deregulation electricity
prices, *106*
Alberta Power, 92
Alberta Power Pool, 4, 94–97, 104, 105
Alberta restructuring experience:
auctions for, 99–104; background
for, 91, 92–93; balancing pool
bidding policy change, *108*; collusive
behavior question in, 108–9;
consumer impact of, 100; consumer
rate options adopted in, 105; early
phases of, 93–106; external
transmission connections impacting,
93; future of, 109–10; generating
plant construction/expansion during,
97–98; generating plant ownership
aspects of, 98–99; generating plants
of interest in, 101; import/export
patterns and, *107*, 108; *legislated*

hedges and, 95–97; legislation
impacting, 93–95; location
characteristics of, 91; long-term
issues implied by, 109–10; low cost
nature of pricing system, 95–96;
marginal-cost pricing in the
wholesale market of, 95–96; market
tightening during, *96*, 97; *Merit
Order of Supply Prior to
Restructuring*, 94; monopoly position
of distributors and, 97–98; outside
influences on, 104–6, 107; ownership
restriction in, 29; political factors
influencing, 106; power pool
established for, 94–97, 104, 105;
Power Purchase Agreements in,
98–106; recent developments in,
106–10; retail price capping during,
105; summarized, 4; supply-side
competition and, 104; support for,
95; uncompetitive behavior question
in, 108–9
antitrust enforcement, 18. *See also*
monopoly
Argentina, 74
Arizona Corporation Commission
(ACC), 64, 192
Arizona utility study, 64–68

About the Contributors

ABOUT THE EDITOR

Andrew N. Kleit is a Research Fellow at the Independent Institute and Professor of Energy, Environmental, and Mineral Economics at Pennsylvania State University. Having received his Ph.D. at the Yale University, Professor Kleit has also held faculty positions at Louisiana State University and Yale University. He has been Junior Staff Economist, President's Council of Economic Advisers; and Economist, Division of Economic Policy Analysis, Bureau of Economics, Federal Trade Commission.

Professor Kleit's books include *Customer Choice, Competition Policy Enforcement: The Economics of the Antitrust Process* (with Malcolm B. Coate) and *Disentangling Regulatory Policy: The Effects of State Regulations on Trucking Rates* (with Timothy Daniel). A contributor to numerous scholarly volumes, his articles have appeared in *Applied Economics, Health Economics, Review of Economics and Statistics, Resource and Energy Economics, Review of Industrial Organization, Journal of Law and Economics, Energy Journal, Economic Inquiry, Journal of Regulatory Economics, Journal of Regulatory Economics, Managerial and Decision Economics, Research in Law and Economics, Southern Economic Journal*, and *Journal of Institutional and Theoretical Economics*.

ABOUT THE AUTHORS

Timothy J. Brennan is professor of policy sciences and economics at the University of Maryland, Baltimore.

Timothy J. Considine is professor of natural resource economics at the Pennsylvania State University.

Terry Daniel is professor of management science at the University of Alberta School of Business.

David E. Dismukes is associate professor and associate director at the Louisiana State University Center for Energy Studies.

Joseph Doucet is Enbridge Professor of Energy Policy at the University of Alberta School of Business.

Michael Giberson is a postdoctoral fellow at the Critical Infrastructure Protection Project and the Interdisciplinary Center for Economic Science, both of George Mason University.

William W. Hogan is Lucius N. Littauer Professor of Public Policy and Administration at the John F. Kennedy School of Government, Harvard University.

L. Lynne Kiesling is director, Center for Applied Energy Research, research scholar, Interdisciplinary Center for Economic Science at George Mason University, and senior lecturer, Department of Economics, Northwestern University.

Craig S. Pirrong is professor of finance at the University of Houston.

André Plourde is professor of economics and chair of the Economics Department at the University of Alberta.

Stephen J. Rassenti is professor of economics at the Interdisciplinary Center for Economic Science at George Mason University.

Vernon L. Smith is professor of economics and law at the Interdisciplinary Center for Economic Science at George Mason University.

Bart J. Wilson is associate professor of economics at the Interdisciplinary Center for Economic Science at George Mason University.